CALC

COSM●S

How Mathematics Unveils the Universe

By the same author

CALCULATING THE COSMOS

How Mathematics Unveils the Universe

IAN STEWART

PROFILE BOOKS

This paperback edition published in 2017

First published in Great Britain in 2016 by
PROFILE BOOKS LTD
3 Holford Yard
Bevin Way
London
WC1X 9HD
www.profilebooks.com

10 9 8 7 6 5 4 3 2 1

Typeset in Sabon by MacGuru Ltd
Printed and bound in Great Britain by CPI Group (UK) Ltd, Croydon CR0 4YY

A CIP catalogue record for this book is available from the British Library.

ISBN 978 1 78125 433 2
eISBN 978 1 78283 150 1

Contents

Prologue

'Why, I have calculated it.'

Isaac Newton's reply to Edmond Halley, when asked how he knew that an inverse square law of attraction implies that the orbit of a planet is an ellipse. Quoted in Herbert Westren Turnbull, *The Great Mathematicians*.

ON 12 NOVEMBER 2014 an intelligent alien observing the solar system would have witnessed a puzzling event. For months, a tiny machine had been following a comet along its path round the Sun – passive, dormant. Suddenly, the machine awoke, and spat out an even tinier machine. This descended towards the coal-black surface of the comet, hit ... and bounced. When it finally came to a halt, it was tipped over on its side and jammed against a cliff.

The alien, deducing that the landing had not gone as intended, might not have been terribly impressed, but the engineers behind the two machines had pulled off an unprecedented feat – landing a space probe on a comet. The larger machine was *Rosetta*, the smaller *Philae*, and the comet was 67P/Churyumov–Gerasimenko. The mission was carried out by the European Space Agency, and the flight alone took more than ten years. Despite the bumpy landing, *Philae* attained most of its scientific objectives and sent back vital data. *Rosetta* continues to perform as planned.

Why land on a comet? Comets are intriguing in their own right, and anything we can discover about them is a useful addition to basic science. On a more practical level, comets occasionally come close to the Earth, and a collision would cause huge devastation, so it's prudent to find out what they're made of. You can change the orbit of a solid body using a rocket or a nuclear missile, but a soft spongy one might break up and make the problem worse. However, there's a third reason. Comets contain material that goes back to the origin of the solar system, so they can provide useful clues about how our world came into being.

Astronomers think that comets are dirty snowballs, ice covered in a thin layer of dust. *Philae* managed to confirm this, for comet 67P, before its batteries discharged and it went silent. If the Earth formed at its current distance from the Sun, it has more water than it ought to. Where did the extra water come from? An attractive possibility is bombardment by millions of comets when the solar system was forming. The ice melted, and the oceans were born. Perhaps surprisingly, there's a way to test this theory. Water is made from hydrogen and oxygen. Hydrogen occurs in three distinct atomic forms, known as isotopes; these all have the same number of protons and electrons (one of each), but differ in the number of neutrons. Ordinary hydrogen has no neutrons, deuterium has one, and tritium has two. If Earth's oceans came from comets, the proportions of those isotopes in the oceans, and in the crust, whose rocks also contain large amounts of water within their chemical make-up, should be similar to their proportions in comets.

Philae's analysis shows that compared to Earth, 67P has a much greater proportion of deuterium. More data from other comets will be needed to make sure, but a cometary origin for the oceans is starting to look shaky. Asteroids are a better bet.

The 'rubber duck' comet 67P, imaged by Rosetta.

✦

The *Rosetta* mission is just one example of humanity's growing ability to send machines into space, either for scientific exploration or for everyday use. This new technology has expanded our scientific aspirations. Our space probes have now visited, and sent back their holiday snaps of, every planet in the solar system, and some of the smaller bodies.

Progress has been rapid. American astronauts landed on the Moon in 1969. The *Pioneer 10* spacecraft, launched in 1972, visited Jupiter and continued out of the solar system. *Pioneer 11* followed it in 1973 and visited Saturn as well. In 1977 *Voyager 1* and *Voyager 2* set out to explore these worlds, and the even more distant planets Uranus and Neptune. Other craft, launched by several different nations or national groupings, have visited Mercury, Venus, and Mars. Some have even *landed* on Venus and Mars, sending back valuable information. As I write in 2015, five orbital probes[1] and two[2] surface vehicles are exploring Mars, *Cassini* is in orbit around Saturn, the *Dawn* spacecraft is orbiting the former asteroid and recently promoted dwarf planet Ceres, and the *New Horizons* spacecraft has just whizzed past, and sent back stunning images of, the most famous dwarf planet in the solar system: Pluto. Its data will help resolve the mysteries of this enigmatic body and its five moons. It has already shown that Pluto is marginally larger than Eris, a more distant dwarf planet previously thought to be the largest. Pluto was reclassified as a dwarf planet in order to exclude Eris from planetary status. Now we discover they needn't have bothered.

We're also starting to explore lesser but equally fascinating bodies: moons, asteroids, and comets. It may not be *Star Trek*, but the final frontier is opening up.

Space exploration is basic science, and while most of us are intrigued by new discoveries about the planets, some prefer their tax contributions to produce more down-to-earth payoffs. As far as everyday life is concerned, our ability to create accurate mathematical models of bodies interacting under gravity has given the world a range of technological wonders that rely on artificial satellites: satellite television, a highly efficient international telephone network, weather satellites, satellites watching the Sun for magnetic storms, satellites keeping

On 14 July 2015 NASA's New Horizons space probe sent this historic image
of Pluto to Earth, the first to show clear features on the dwarf planet.

watch on the environment and mapping the globe – even car satnav,
using the Global Positioning System.

These accomplishments would have amazed previous generations.
Even in the 1930s, most people thought that no human would ever
stand on the Moon. (Today plenty of rather naive conspiracy theorists
still think nobody has, but don't get me started.) There were heated
arguments about even the bare *possibility* of spaceflight.[3] Some people
insisted that rockets wouldn't work in space because 'there is nothing
there to push against', unaware of Newton's third law of motion – to
every action there is an equal and opposite reaction.[4]

Serious scientists stoutly insisted that a rocket would never work
because you needed a lot of fuel to lift the rocket, then more fuel to lift
the fuel, then more fuel to lift *that*… even when a picture in the four-
teenth-century Chinese *Huolongjing* (Fire Dragon Manual) by Jiao Yu
depicts a fire-dragon, aka multistage rocket. This Chinese naval weapon
used discardable boosters to launch an upper stage shaped like the head
of a dragon, loaded with fire arrows that shot out of its mouth. Conrad
Haas made the first European experiment with multistage rockets in
1551. Twentieth-century rocketry pioneers pointed out that the first
stage of a multistage rocket would be able to lift the second stage

and its fuel, while *dropping off* all of the excess weight of the now-exhausted first stage. Konstantin Tsiolkovsky published detailed and realistic calculations about the exploration of the solar system in 1911.

Well, we got to the Moon despite the naysayers – using precisely the ideas that they were too blinkered to contemplate. So far, we've explored only our local region of space, which pales into insignificance compared to the vast reaches of the universe. We've not yet landed humans on another planet, and even the nearest star seems utterly out of reach. With existing technology, it would take centuries to get there, even if we could build a reliable starship. But we're on our way.

✦

These advances in space exploration and usage depend not just on clever technology, but also on a lengthy series of scientific discoveries that go back at least as far as ancient Babylon three millennia ago. Mathematics lies at the heart of these advances. Engineering is of course vital too, and discoveries in many other scientific disciplines were needed before we could make the necessary materials and assemble them into a working space probe, but I'll concentrate on how mathematics has improved our knowledge of the universe.

The story of space exploration and the story of mathematics have gone hand in hand from the earliest times. Mathematics has proved essential for understanding the Sun, Moon, planets, stars, and the vast panoply of associated objects that together form the cosmos – the universe considered on a grand scale. For thousands of years, mathematics has been our most effective method of understanding, recording, and predicting cosmic events. Indeed in some cultures, such as ancient India around 500, mathematics was a sub-branch of astronomy. Conversely, astronomical phenomena have influenced the development of mathematics for over three millennia, inspiring everything from Babylonian predictions of eclipses to calculus, chaos, and the curvature of spacetime.

Initially, the main astronomical role of mathematics was to record observations and perform useful calculations about phenomena such as solar eclipses, where the Moon temporarily obscures the Sun, or lunar eclipses, where the Earth's shadow obscures the Moon. By thinking about the geometry of the solar system, astronomical pioneers realised that the Earth goes round the Sun, even though it looks the other way round from down here. The ancients also combined observations with geometry to estimate the size of the Earth and the distances to the Moon and the Sun.

Deeper astronomical patterns began to emerge around 1600, when Johannes Kepler discovered three mathematical regularities – 'laws' – in the orbits of the planets. In 1679 Isaac Newton reinterpreted Kepler's laws to formulate an ambitious theory that described not just how the planets of the solar system move, but the motion of *any* system of celestial bodies. This was his theory of gravity, one of the central discoveries in his world-changing *Philosophiae Naturalis Principia Mathematica* (Mathematical Principles of Natural Philosophy). Newton's law of gravity describes how each body in the universe attracts every other body.

By combining gravity with other mathematical laws about the motion of bodies, pioneered by Galileo a century earlier, Newton explained and predicted numerous celestial phenomena. More generally, he changed how we think about the natural world, creating a scientific revolution that is still powering ahead today. Newton showed that natural phenomena are (often) governed by mathematical patterns, and by understanding these patterns we can improve our understanding of nature. In Newton's era the mathematical laws explained what was happening in the heavens, but they had no significant practical uses, other than for navigation.

✦

All that changed when the USSR's *Sputnik* satellite went into low Earth orbit in 1957, firing the starting gun for the space race. If you watch football on satellite television – or opera or comedies or science documentaries – you're reaping a real-world benefit from Newton's insights.

Initially, his successes led to a view of the cosmos as a clockwork universe, in which everything majestically follows paths laid down at the dawn of creation. For example, it was believed that the solar system was created in pretty much its current state, with the same planets moving along the same near-circular orbits. Admittedly, everything jiggled around a bit; the period's advances in astronomical observations had made that abundantly clear. But there was a widespread belief that nothing had changed, did change, or would change in any dramatic manner over countless eons. In European religion it was unthinkable that God's perfect creation could have been different in the past. The mechanistic view of a regular, predictable cosmos persisted for three hundred years.

No longer. Recent innovations in mathematics, such as chaos theory, coupled to today's powerful computers, able to crunch the relevant numbers with unprecedented speed, have greatly changed our views of the cosmos. The clockwork model of the solar system remains valid over short periods of time, and in astronomy a million years is usually short. But our cosmic backyard is now revealed as a place where worlds did, and will, migrate from one orbit to another. Yes, there are very long periods of regular behaviour, but from time to time they are punctuated by bursts of wild activity. The immutable laws that gave rise to the notion of a clockwork universe can also cause sudden changes and highly erratic behaviour.

The scenarios that astronomers now envisage are often dramatic. During the formation of the solar system, for instance, entire worlds collided with apocalyptic consequences. One day, in the distant future, they will probably do so again: there's a small chance that either Mercury or Venus is doomed, but we don't know which. It could be both, and they could take us with them. One such collision probably led to the formation of the Moon. It sounds like something out of science fiction, and it is ... but the best kind, 'hard' science fiction in which only the fantastic new invention goes beyond known science. Except that here there is no fantastic invention, just an unexpected mathematical discovery.

Mathematics has informed our understanding of the cosmos on every scale: the origin and motion of the Moon, the movements and form of the planets and their companion moons, the intricacies of asteroids, comets, and Kuiper belt objects, and the ponderous celestial dance of the entire solar system. It has taught us how interactions with Jupiter can fling asteroids towards Mars, and thence the Earth; why Saturn is not alone in possessing rings; how its rings formed to begin with and why they behave as they do, with braids, ripples, and strange rotating 'spokes'. It has shown us how a planet's rings can spit out moons, one at a time.

Clockwork has given way to fireworks.

✦

From a cosmic viewpoint, the solar system is merely one insignificant bunch of rocks among quadrillions. When we contemplate the universe on a grander scale, mathematics plays an even more crucial role. Experiments are seldom possible and direct observations are difficult, so we have to make indirect inferences instead. People with an anti-science agenda often attack this feature as some kind of weakness. Actually, one of the great strengths of science is the ability to infer things that we can't observe directly from those that we can. The existence of atoms was conclusively established long before ingenious microscopes allowed us to see them, and even then 'seeing' depends on a series of inferences about how the images concerned are formed.

Mathematics is a powerful inference engine: it lets us deduce the *consequences* of alternative hypotheses by pursuing their logical implications. When coupled with nuclear physics – itself highly mathematical – it helps to explain the dynamics of stars, with their many types, their different chemical and nuclear constitutions, their writhing magnetic fields and dark sunspots. It provides insight into the tendency of stars to cluster into vast galaxies, separated by even vaster voids, and explains why galaxies have such interesting shapes. It tells us why galaxies combine to form galactic clusters, separated by even vaster voids.

There's a yet larger scale, that of the universe as a whole. This is the realm of cosmology. Here humanity's source of rational inspiration is almost entirely mathematical. We can observe some aspects of the universe, but we can't experiment on it as a whole. Mathematics helps us to interpret observations, by permitting 'what if' comparisons between alternative theories. But even here, the starting point was closer to home. Albert Einstein's general theory of relativity, in which the force of gravity is replaced by the curvature of spacetime, replaced Newtonian physics. The ancient geometers and philosophers would have approved: dynamics was reduced to geometry. Einstein saw his theories verified by two of his own predictions: known, but puzzling, changes to the orbit of Mercury, and the bending of light by the Sun, observed during a solar eclipse in 1919. But he couldn't have realised that his theory would lead to the discovery of some of the most bizarre objects in the entire universe: black holes, so massive that even light can't escape their gravitational pull.

He certainly failed to recognise one potential consequence of his theory, the Big Bang. This is the proposal that the universe originated from a single point at some time in the distant past, around 13·8 billion years ago according to current estimates, in a sort of gigantic explosion. But it was spacetime that exploded, not something else exploding within spacetime. The first evidence for this theory was Edwin Hubble's discovery that the universe is expanding. Run everything backwards and it all collapses to a point; now restart time in the normal direction to get back to here and now.

Einstein lamented that he could have predicted this, if he'd believed his own equations. That's why we can be confident that he didn't expect it.

In science, new answers open up new mysteries. One of the greatest is dark matter, a completely new kind of matter that seems to be required to reconcile observations of how galaxies spin with our understanding of gravity. However, searches for dark matter have consistently failed to detect any. Moreover, two other add-ons to the original Big Bang theory are also required to make sense of the cosmos. One is inflation, an effect that caused the early universe to grow by a truly enormous amount in a truly tiny instant of time. It's needed to explain why the distribution of matter in today's universe is almost, but not quite, uniform. The other is dark energy, a mysterious force that causes the universe to expand at an ever-faster rate.

The Big Bang is accepted by most cosmologists, but only when these three extras – dark matter, inflation, and dark energy – are thrown in to the mix. However, as we shall see, each of these *dei ex machina* comes with a host of troubling problems of its own. Modern cosmology no longer seems as secure as it was a decade ago, and there might be a revolution on the way.

✦

Newton's law of gravity wasn't the first mathematical pattern to be discerned in the heavens, but it crystallised the whole approach, as well as going far beyond anything that had come before. It's a core theme of *Calculating the Cosmos*, a key discovery that lies at the heart of the book. Namely: there are mathematical patterns in the motions and structure of both celestial and terrestrial bodies, from the smallest dust particle to the universe as a whole. Understanding those patterns allows us not just to explain the cosmos, but also to explore it, exploit it, and protect ourselves against it.

Arguably the greatest breakthrough is to realise that there *are* patterns. After that, you know what to look for, and while it may be difficult to pin the answers down, the problems become a matter of technique. Entirely new mathematical ideas often have to be invented – I'm not claiming it's easy or straightforward. It's a long-running game and it's still playing out.

Newton's approach also triggered a standard reflex. As soon as the latest discovery hatches from its shell, mathematicians start wondering whether a similar idea might solve other problems. The urge to make everything more general runs deep in the mathematical psyche. It's no good blaming it on Nicolas Bourbaki[5] and the 'new maths': it goes back to Euclid and Pythagoras. From this reflex, mathematical physics was born. Newton's contemporaries, mainly in continental Europe, applied the same principles that had plumbed the cosmos to understand heat, sound, light, elasticity, and later electricity and magnetism. And the message rang out ever clearer:

Nature has laws.
They are mathematical.
We can find them.
We can use them.

Of course, it wasn't that simple.

Attraction at a Distance

Macavity, Macavity, there's no one like Macavity,
He's broken every human law, he breaks the law of gravity.
Thomas Stearns Eliot, *Old Possum's Book of Practical Cats*

WHY DO THINGS FALL down?

Some don't. Macavity, obviously. Along with the Sun, the Moon, and almost everything else 'up there' in the heavens. Though rocks sometimes fall from the sky, as the dinosaurs discovered to their dismay. Down here, if you want to be picky, insects, birds, and bats fly, but they don't stay up indefinitely. Pretty much everything else falls, unless something is holding it up. But up there, nothing holds it up – yet it doesn't fall.

Up there seems very different from down here.

It took a stroke of genius to realise that what makes terrestrial objects fall is the same thing that holds celestial objects up. Newton famously compared a falling apple to the Moon, and realised that the Moon stays up because, unlike the apple, it's also moving *sideways*.[1] Actually, the Moon is perpetually falling, but the Earth's surface falls away from it at the same rate. So the Moon can fall forever, yet go round and round the Earth and never hit it.

The real difference was not that apples fall and Moons don't. It was that apples don't move sideways fast enough to miss the Earth.

Newton was a mathematician (and a physicist, chemist, and mystic), so he did some sums to confirm this radical idea. He calculated the forces that must be acting on the apple and the Moon to make them follow their separate paths. Taking their different masses into account, the forces turned out to be identical. This convinced him that the Earth must be pulling both apple and Moon towards it. It was natural to suppose that the same type of attraction holds for any pair of bodies, terrestrial or celestial. Newton expressed those attractive forces in a mathematical equation, a law of nature.

One remarkable consequence is that not only does the Earth attract the apple: the apple also attracts the Earth. And the Moon, and everything else in the universe. But the apple's effect on the Earth is way too small to measure, unlike the Earth's effect on the apple.

This discovery was a huge triumph, a deep and precise link between mathematics and the natural world. It also had another important implication, easily missed among the mathematical technicalities: despite appearances, 'up there' is in some vital respects the same as 'down here'. The laws are identical. What differs is the context in which they apply.

We call Newton's mysterious force 'gravity'. We can calculate its effects with exquisite accuracy. We still don't understand it.

✦

For a long time, we thought we did. Around 350 BC the Greek philosopher Aristotle gave a simple reason why objects fall down: they are seeking their natural resting place.

To avoid circular reasoning, he also explained what 'natural' meant. He maintained that everything is made from four basic elements: earth, water, air, and fire. The natural resting place of earth and water are at the centre of the universe, which of course coincides with the centre of the Earth. As proof, the Earth doesn't move: we live on it, and would surely notice if it did. Since earth is heavier than water (it sinks, right?) the lowest regions are occupied by earth, a sphere. Next comes a spherical shell of water, then one of air (air is lighter than water: bubbles rise). Above that – but lower than the celestial sphere that carries the Moon – is the realm of fire. All other bodies tend to rise or fall according to the proportions in which these four elements occur.

This theory led Aristotle to argue that the speed of a falling body is proportional to its weight (feathers fall more slowly than stones) and inversely proportional to the density of the surrounding medium (stones fall faster in air than in water). Having reached its natural rest state, the body remains there, moving only when a force is applied.

As theories go, these aren't so bad. In particular, they agree with everyday experience. On my desk, as I write, there is a first edition of the novel *Triplanetary*, quoted in the epigram for Chapter 2. If I leave it alone, it stays where it is. If I apply a force – give it a shove – it moves a few centimetres, slowing down as it does so, and stops.

Aristotle was right.

And so it seemed for nigh on two thousand years. Aristotelian physics, though widely debated, was generally accepted by almost all intellectuals until the end of the sixteenth century. An exception was the Arab scholar al-Hasan ibn al-Haytham (Alhazen), who argued against Aristotle's view on geometric grounds in the eleventh century. But even today, Aristotelian physics matches our intuition more closely than do the ideas of Galileo and Newton that replaced it.

To modern thinking, Aristotle's theory has some big gaps. One is weight. *Why* is a feather lighter than a stone? Another is friction. Suppose I placed my copy of *Triplanetary* on an ice-skating rink and gave it the same push. What would happen? It would go further: a lot further if I rested it on a pair of skates. Friction makes a body move more slowly in a viscous – sticky – medium. In everyday life, friction is everywhere, and that's why Aristotelian physics matches our intuition better than Galilean and Newtonian physics do. Our brains have evolved an internal model of motion with friction built in.

Now we know that a body falls towards the Earth because the planet's gravity pulls it. But what is gravity? Newton thought it was a force, but he didn't explain how the force arose. It just *was*. It acted at a distance without anything in between. He didn't explain how it did that either; it just *did*. Einstein replaced force by the curvature of spacetime, making 'action at a distance' irrelevant, and he wrote down equations for how curvature is affected by a distribution of matter – but he didn't explain *why* curvature behaves like that.

People calculated aspects of the cosmos, such as eclipses, for millennia before anyone realised that gravity existed. But once gravity's role was revealed, our ability to calculate the cosmos became far more powerful. Newton's subtitle for Book 3 of the *Principia*, which described his laws of motion and gravity, was 'Of the System of the World'. It was only a slight exaggeration. The force of gravity, and the manner in which bodies respond to forces, lie at the heart of most cosmic calculations. So before we get to the latest discoveries, such as how ringed planets spit out moons, or how the universe began, we'd better sort out some basic ideas about gravity.

✦

Before the invention of street lighting, the Moon and stars were as familiar, to most people, as rivers, trees, and mountains. As the Sun went down, the stars came out. The Moon marched to its own drummer, sometimes appearing during the day as a pale ghost, but shining much more brightly at night. Yet there were patterns. Anyone observing the Moon even casually for a few months would quickly notice that it follows a regular rhythm, changing shape from a thin crescent to a circular disc and back again every 28 days. It also moves noticeably from one night to the next, tracing a closed, repetitive path across the heavens.

The stars have their own rhythm too. They revolve, once a day, round a fixed point in the sky, as if they're painted on the inside of a slowly spinning bowl. *Genesis* talks of the firmament of Heaven: the Hebrew word translated as 'firmament' means bowl.

Observing the sky for a few months, it also became obvious that five stars, including some of the brightest, don't revolve like the majority of 'fixed' stars. Instead of being attached to the bowl, they crawl slowly across it. The Greeks associated these errant specks of light with Hermes (messenger of the gods), Aphrodite (goddess of love), Ares (god of war), Zeus (king of the gods), and Kronos (god of agriculture). The corresponding Roman deities gave them their current English names: Mercury, Venus, Mars, Jupiter, and Saturn. The Greeks called them *planetes*, 'wanderers', hence the modern name planets, of which we now recognise three more: Earth, Uranus, and Neptune. Their paths were strange, seemingly unpredictable. Some moved relatively quickly, others were slower. Some even looped back on themselves as the months passed.

Most people just accepted the lights for what they were, in the same way that they accepted the existence of rivers, trees, and mountains. But a few asked questions. What are these lights? Why are they there? How and why do they move? Why do some movements show patterns, while others break them?

The Sumerians and Babylonians provided basic observational data. They wrote on clay tablets in a script known as cuneiform – wedge-shaped. Among the Babylonian tablets that archaeologists have found are star catalogues, listing the positions of stars in the sky; they date to about 1200 BC but were probably copies of even earlier Sumerian tablets. The Greek philosophers and geometers who followed their lead were more aware of the need for logic, proof, and theory. They were pattern-seekers; the Pythagorean cult took this attitude to extremes, believing that the entire universe is run by numbers. Today most scientists would agree, but not about the details.

The Greek geometer who had the most influence on the astronomical thinking of later generations was Claudius Ptolemy, an astronomer and geographer. His earliest work is known as the *Almagest*, from an Arabic rendering of its original title, which started out as 'The Mathematical Compilation', morphed into 'The Great Compilation', and then into '*al-majisti*' – the greatest. The *Almagest* presented a fully fledged theory of planetary motion based on what the Greeks considered to be the most perfect of geometric forms, circles and spheres.

The planets do not, in fact, move in circles. This wouldn't have been news to the Babylonians, because it doesn't match their tables. The Greeks went further, asking what would match. Ptolemy's answer was: combinations of circles supported by spheres. The innermost sphere, the 'deferent', is centred on the Earth. The axis of the second sphere, or 'epicycle', is fixed to the sphere just inside it. Each pair of spheres is disconnected from the others. It wasn't a new idea. Two centuries earlier, Aristotle – building on even earlier ideas of the same kind – had proposed a complex system of 55 concentric spheres, with the axis of each sphere fixed to the sphere just inside it. Ptolemy's modification used fewer spheres, and was more accurate, but it was still rather complicated. Both led to the question whether the spheres actually existed, or were just convenient fictions – or whether something entirely different was really going on.

✦

For the next thousand years and more, Europe turned to matters theological and philosophical, basing most of its understanding of the natural world on what Aristotle had said around 350 BC. The universe was believed to be geocentric, with everything revolving around a stationary Earth. The torch of innovation in astronomy and mathematics passed to Arabia, India, and China. With the dawn of the Italian Renaissance, however, the torch passed back to Europe. Subsequently, three giants of science played leading roles in the advance of astronomical knowledge: Galileo, Kepler, and Newton. The supporting cast was huge.

Galileo is famous for his invention of improvements to the telescope, with which he discovered that the Sun has spots, Jupiter has (at least) four moons, Venus has phases like the Moon's, and there's something strange about Saturn – later explained as its ring system. This evidence led him to reject the geocentric theory and embrace Nicolaus Copernicus's rival heliocentric theory, in which the planets and the Earth revolve round the Sun, getting Galileo into trouble with the Church of Rome. But he also made an apparently more modest, but ultimately more important, discovery: a mathematical pattern in the motion of objects such as cannonballs. Down here, a freely moving body either speeds up (when falling) or slows down (when rising) by an amount that is the same over a fixed, *small* period of time. In short, the body's acceleration is constant. Lacking accurate clocks, Galileo observed these effects by rolling balls down gentle inclines.

The next key figure is Kepler. His boss Tycho Brahe had made very accurate measurements of the position of Mars. When Tycho died, Kepler inherited his position as astronomer to Holy Roman Emperor Rudolph II, together with his observations, and set about calculating the true shape of Mars's orbit. After fifty failures, he deduced that the orbit is shaped like an ellipse – an oval, like a squashed circle. The Sun lies at a special point, the focus of the ellipse.

Ellipses were familiar to the ancient Greek geometers, who defined them as plane sections of a cone. Depending on the angle of the plane relative to the cone, these 'conic sections' include circles, ellipses, parabolas, and hyperbolas.

When a planet moves in an ellipse, its distance from the Sun varies. When it comes close to the Sun, it speeds up; when it's more distant, it slows down. It's a bit of a surprise that these effects conspire to create an orbit that has exactly the same shape at both ends. Kepler didn't expect this, and for a long time it persuaded him that an ellipse must be the wrong answer.

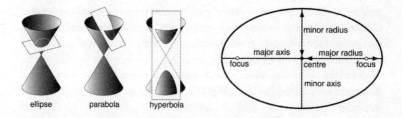

Left: Conic sections. Right: Basic features of an ellipse.

The shape and size of an ellipse are determined by two lengths: its major axis, which is the longest line between two points on the ellipse, and its minor axis, which is perpendicular to the major axis. A circle is a special type of ellipse for which these two distances are equal; they then give the diameter of the circle. For astronomical purposes the radius is a more natural measure – the radius of a circular orbit is the planet's distance from the Sun – and the corresponding quantities for an ellipse are called the major radius and minor radius. These are often referred to by the awkward terms semi-major axis and semi-minor axis, because they cut the axes in half. Less intuitive but very important is the eccentricity of the ellipse, which quantifies how long and thin it is. The eccentricity is o for a circle and for a fixed major radius it becomes infinitely large as the minor radius tends to zero.[2]

The size and shape of an elliptical orbit can be characterised by two numbers. The usual choice is the major radius and the eccentricity. The minor radius can be found from these. The Earth's orbit has major radius 149·6 million kilometres and eccentricity 0·0167. The minor radius is 149·58 million kilometres, so the orbit is very close to a circle, as the small eccentricity indicates. The plane of the Earth's orbit has a special name: the ecliptic.

The spatial location of any other elliptical orbit about the Sun can be characterised by three more numbers, all angles. One is the inclination of the orbital plane to the ecliptic. The second effectively gives the direction of the major axis in that plane. The third gives the direction of the line at which the two planes meet. Finally, we need to know where the planet is in the orbit, which requires one further angle. So specifying the orbit of the planet and its position within that orbit requires two numbers and four angles – six *orbital elements*. A major goal of early astronomy was to calculate the orbital elements of every planet and asteroid that was discovered. Given these numbers, you can predict its future motion, at least until the combined effects of the other bodies disturb its orbit significantly.

Kepler eventually came up with a set of three elegant mathematical patterns, now called his laws of planetary motion. The first states that the orbit of a planet is an ellipse with the Sun at one focus. The second says that the line from the Sun to the planet sweeps out equal areas in equal periods of time. And the third tells us that the square of the period of revolution is proportional to the cube of the distance.

✦

Newton reformulated Galileo's observations about freely moving bodies as three laws of motion. The first states that bodies continue to move in a straight line at a constant speed unless acted on by a force. The second states that the acceleration of any body, multiplied by its mass, is equal to the force acting on it. The third states that every action produces an equal and opposite reaction. In 1687 he reformulated Kepler's planetary laws as a general rule for how heavenly bodies move – the law of gravity, a mathematical formula for the gravitational force with which any body attracts any other.

Indeed, he *deduced* his force law from Kepler's laws by making one assumption: the Sun exerts an attractive force, always directed towards its centre. On this assumption, Newton proved that the force is inversely proportional to the square of the distance. That's a fancy way to say that, for example, multiplying the mass of either body by three also trebles the force, but multiplying the distance between them by three reduces the force to one ninth of the amount. Newton also proved the converse: this 'inverse square law' of attraction implies Kepler's three laws.

Credit for the law of gravity rightly goes to Newton, but the idea wasn't original with him. Kepler deduced something similar by analogy with light, but thought gravity pushed planets round their orbits. Ismaël Bullialdus disagreed, arguing that the force of gravity must be inversely proportional to the square of the distance. In a lecture to the Royal Society in 1666, Robert Hooke said that that all bodies move in a straight line unless acted on by a force, all bodies attract each other gravitationally, and the force of gravity decreases with distance by a formula that 'I own I have not discovered'. In 1679 he settled on an inverse square law for the attraction, and wrote to Newton about it.[3] So Hooke was distinctly miffed when exactly the same thing appeared in *Principia*, even though Newton credited him, along with Halley and Christopher Wren.

Hooke did accept that only Newton had deduced that closed orbits are elliptical. Newton knew that the inverse square law also permits parabolic and hyperbolic orbits, but these aren't closed curves, so the motion doesn't repeat periodically. Orbits of those kinds also have astronomical applications, mainly to comets.

Newton's law goes beyond Kepler's because of one further feature, a prediction rather than a theorem. Newton realised that since the Earth attracts the Moon, it seems reasonable that the Moon should also attract the Earth. They're like two country dancers, holding hands and whirling round and round. Each dancer feels the force exerted by the other, tugging at their arms. Each dancer is held in place by that force: if they let go, they will spin off across the dance floor. However, the Earth is much more massive than the Moon, so it's like a fat man dancing with a small child. The man seems to spin in place as the child whirls round and round. But look carefully, and you'll see that the fat man is whirling too: his feet go round in a small circle, and the centre about which he rotates is slightly closer to the child than it would have been if he were spinning alone.

This reasoning led Newton to propose that *every* body in the universe attracts every other body. Kepler's laws apply to only two bodies, Sun and planet. Newton's law applies to any system of bodies whatsoever, because it provides both the magnitude and the direction of *all of the forces that occur*. Inserted into the laws of motion, the combination of all these forces determines each body's acceleration, hence velocity, hence position at any moment. The enunciation of a universal law of gravity was an epic moment in the history and development of science, revealing hidden mathematical machinery that keeps the universe ticking.

✦

Newton's laws of motion and gravity triggered a lasting alliance between astronomy and mathematics, leading to much of what we now know about the cosmos. But even when you understand what the laws are, it's not straightforward to apply them to specific problems. The gravitational force, in particular, is 'nonlinear', a technical term whose main implication is that you can't solve the equations of motion using nice formulas. Or nasty ones, for that matter.

Post-Newton, mathematicians got round this obstacle either by working with very artificial (though intriguing) problems, such as three identical masses arranged in an equilateral triangle, or by deriving approximate solutions to more realistic problems. The second approach is more practical, but actually a lot of useful ideas came from the first, artificial though it was.

For a long time, Newton's scientific heirs had to perform their calculations by hand, often a heroic task. An extreme example is Charles-Eugène Delaunay, who in 1846 started to calculate an approximate formula for the motion of the Moon. The feat took over twenty years, and he published his results in two volumes. Each has more than 900 pages, and the second volume consists entirely of the formula. In the late twentieth century his answer was checked using computer algebra (software systems that can manipulate formulas, not just numbers). Only two tiny errors were found, one a consequence of the other. Both have a negligible effect.

The laws of motion and gravity are of a special kind, called differential equations. Such equations specify the rate at which quantities change as time passes. Velocity is the rate of change of position, acceleration the rate of change of velocity. The rate at which a quantity is currently changing lets you project its value into the future. If a car is travelling at ten metres per second, then one second from now it will have moved ten metres. This type of calculation requires the rate of change to be constant, however. If the car is accelerating, then one second from now it will have moved more than ten metres. Differential equations get round this problem by specifying the instantaneous rate of change. In effect, they work with very small intervals of time, so that the rate of change can be considered constant during that time interval. It actually took mathematicians several hundred years to make sense of that idea in full logical rigour, because no finite period of time can be instantaneous unless it's zero, and nothing changes in zero time.

Computers created a methodological revolution. Instead of

calculating approximate formulas for the motion, and then putting the numbers into the formulas, you can work from the beginning with the numbers. Suppose you want to predict where some system of bodies – say the moons of Jupiter – will be in a hundred years' time. Start from the initial positions and motions of Jupiter, its moons, and any other bodies that might be important, such as the Sun and Saturn. Then, tiny time step by tiny time step, compute how the numbers describing *all* the bodies change. Repeat until you reach a hundred years: stop. A human with pencil and paper couldn't use this method on any realistic problem. It would take lifetimes. With a fast computer, however, the method becomes entirely feasible. And modern computers are very fast indeed.

It's not *quite* that easy, to be honest. Although the error at each step (caused by assuming a constant rate of change when actually it varies a little) is very small, you have to use an awful lot of steps. A big number times a small error need not be small, but carefully concocted methods keep the errors under control. The branch of mathematics known as numerical analysis is aimed at just this issue. It's convenient to refer to such methods as 'simulations', reflecting the crucial role of the computer. It's important to appreciate that you can't solve a problem merely by 'putting it on the computer'. Someone has to program the machine with the mathematical rules that make its computations match reality.

So exquisitely accurate are those rules that astronomers can predict eclipses of the Sun and Moon to the second, and predict within a few kilometres whereabouts on the planet they will occur, hundreds of years into the future. These 'predictions' can also be run *backwards* in time to pin down exactly when and where historically recorded eclipses occurred. These data have been used to date observations made thousands of years ago by Chinese astronomers, for example.

✦

Even today, mathematicians and physicists are discovering new and unexpected consequences of Newton's law of gravity. In 1993 Cris Moore used numerical methods to show that three bodies with identical masses can chase each other repeatedly along the same figure-8 shaped orbit, and in 2000 Carles Simó showed numerically that this orbit is stable, except perhaps for a slow drift. In 2001 Alain Chenciner and Richard Montgomery gave a rigorous proof that this orbit exists, based on the principle of least action, a fundamental theorem in classical mechanics.[4] Simó has discovered many similar 'choreographies', in which several bodies with the same mass pursue each other along exactly the same (complicated) path.[5]

The stability of the figure-8 three-body orbit seems to persist if the masses are slightly different, opening up a small possibility that three real stars might behave in this remarkable way. Douglas Heggie estimates that there might be one triple system of this kind per galaxy, and there's a fair chance of at least one somewhere in the universe.

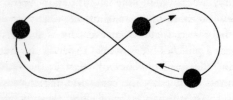

The Fig-8 three-body orbit.

These orbits all exist in a plane, but there's a novel three-dimensional possibility. In 2015 Eugene Oks realised that unusual orbits of electrons in 'Rydberg quasimolecules' might also occur in Newtonian gravity. He showed that a planet can be batted back and forth between the two stars of a binary system in a corkscrew orbit that spirals around the line that joins them.[6] The spirals are loose in the middle but tighten up near the stars. Think of joining the stars by a rotating slinky, stretched in the middle and doubling back on itself at the ends. For stars with different masses, the slinky should be tapered like a cone. Orbits like this can be stable, even if the stars don't move in circles.

Collapsing gas clouds create planar orbits, so a planet is unlikely to form in a such an orbit. But a planet or asteroid perturbed into in a highly tilted orbit might rarely be captured by the binary stars and end up corkscrewing between them. There's tentative evidence that Kepler-16b, a planet orbiting a distant star, might be one of them.

✦

One aspect of Newton's law bothered the great man himself; in fact, it bothered him more than it did most of those who built on his work. The law describes the force that one body exerts on another, but it does not address *how* the force works. It postulates a mysterious 'action at a distance'. When the Sun attracts the Earth, somehow the Earth must 'know' how far it is from the Sun. If, for example, some kind of elastic string joined the two, then the string could propagate the force, and the physics of the string would govern how strong the force was. But between Sun and Earth is only empty space. How does the Sun know how hard to pull the Earth – or the Earth know how hard to be pulled?[7]

Pragmatically, we can apply the law of gravity without worrying about a physical mechanism to transmit the force from one body to another. On the whole, that's what everyone did. A few scientists, however, possess a philosophical streak, a spectacular example being Albert Einstein. His special theory of relativity, published in 1905, changed physicists' view of space, time, and matter. Its extension in 1915 to general relativity changed their view of gravity, and almost as a side issue resolved the thorny question of how a force could act at a distance. It did so by getting rid of the force.

Einstein deduced special relativity from a single fundamental principle: the speed of light remains unchanged even when the observer is moving at a constant speed. In Newtonian mechanics, if you're in an open-top car and you throw a ball in the direction the car is moving, then the speed of the ball as measured by a stationary observer at the roadside will be the speed of the ball relative to the car, *plus* the speed of the car. Similarly, if you shine a torch beam ahead of the car, the speed of light as measured by someone at the side of the road ought to be its usual speed plus that of the car.

Experimental data and some thought experiments persuaded Einstein that light *isn't* like that. The observed speed of light is *the same* for the person shining the torch, and the one at the roadside. The logical consequences of this principle – which I've always felt should be called *non*-relativity – are startling. Nothing can travel faster than light.[8] As a body approaches the speed of light, it shrinks in the direction of motion, its mass increases, and time passes ever more slowly. At the speed of light – if that were possible – it would be infinitely thin, have infinite mass, and time on it would stop. Mass and energy are related: energy equals mass times the square of the speed of light. Finally, events that one observer considers to happen at the same time may not be simultaneous for another observer moving at a constant relative speed to the first.

In Newtonian mechanics, none of these weird things happens. Space is space and time is time, and never the twain shall meet. In special relativity, space and time are to some extent interchangeable, the extent being limited by the speed of light. Together they form a single spacetime continuum. Despite its strange predictions, special relativity has become accepted as the most accurate theory of space and time that we have. Most of its wilder effects only become apparent when objects are travelling very fast, which is why we don't notice them in everyday life.

The most obvious missing ingredient is gravity. Einstein spent years trying to incorporate the force of gravity into relativity, motivated in part by an anomaly in the orbit of Mercury.[9] The end result was general relativity, which extends the formulation of special relativity from a 'flat' spacetime continuum to a 'curved' one. We can get a rough understanding of what's involved by cutting space down to two dimensions instead of three. Now space becomes a plane, and special relativity describes the motion of particles in this plane. In the absence of gravity, they follow straight lines. As Euclid pointed out, a straight line is the shortest distance between two points. To put gravity into the picture, place a star in the plane. Particles no longer follow straight lines; instead, they orbit the star along curves, such as ellipses.

In Newtonian physics, these paths are curved because a force diverts the particle from a straight line. In general relativity, a similar effect is obtained by bending spacetime. Suppose the star distorts the plane, creating a circular valley – a 'gravity well' with the star at the bottom – and assume that moving particles follow whichever path is shortest. The technical term is *geodesic*. Since the spacetime continuum is bent, geodesics are no longer straight lines. For example, a particle can be trapped in the valley, going round and round at a fixed height, like a planet in a closed orbit.

Instead of a hypothetical force that causes the particle's path to curve, Einstein substituted a spacetime that's *already* curved, and whose curvature affects the path of a moving particle. No action at

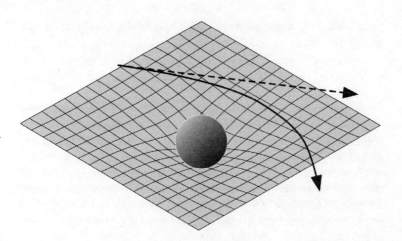

Effect of curvature/gravity on a particle passing a star or planet.

a distance is needed: spacetime is curved because that's what stars do to it, and orbiting bodies respond to nearby curvature. What we and Newton refer to as gravity, and think of as a force, is actually the curvature of spacetime.

Einstein wrote down mathematical formulas, the Einstein field equations,[10] which describe how curvature affects the motion of masses, and how the mass distribution affects curvature. In the absence of any masses, the formula reduces to special relativity. So all of the weird effects, such as time slowing down, also happen in general relativity. Indeed, gravity can *cause* time to slow down, even for an object that is not moving. Usually these paradoxical effects are small, but in extreme circumstances the behaviour that relativity (of either type) predicts differs significantly from Newtonian physics.

Think this all sounds mad? Many did, to begin with. But anyone who uses satnav in their car is relying on both special and general relativity. The calculations that tell you you're on the outskirts of Bristol heading south on the M32 motorway rely on timing signals from orbiting satellites. The chip in your car that computes your location has to correct those timings for two effects: the speed with which the satellite is moving, and its position in Earth's gravity well. The first requires special relativity, the second general relativity. Without these corrections, within a few days the satnav would place you in the middle of the Atlantic.

✦

General relativity shows that Newtonian physics is *not* the true, exact 'system of the world' that he (and almost all other scientists prior to the twentieth century) believed it to be. However, that discovery did not spell the end of Newtonian physics. In fact, it's far more widely used now, and for more practical purposes, than it was in Newton's day. Newtonian physics is simpler than relativity, and it's 'good enough for government work', as they say: literally so. The differences between the two theories mainly become apparent when considering exotic phenomena such as black holes. Astronomers and space mission engineers, mainly employed by governments or contracts between governments and organisations such as NASA and ESA, still use Newtonian mechanics for almost all calculations. There are a few exceptions where timing is delicate. As the story unfolds, we'll see the influence of Newton's law of gravity over and over again. It really is that important: one of the greatest scientific discoveries of all time.

However, when it comes to cosmology – the study of the entire universe and especially its origins – we must ditch Newtonian physics. It can't explain the key observations. Instead, general relativity must be invoked, ably assisted by quantum mechanics. And even those two great theories seem to need extra help.

Collapse of the Solar Nebula

> Two thousand million years or so ago two galaxies were colliding,
> or rather, were passing through each other... At about the same
> time – within the same plus-or-minus ten percent margin of error,
> it is believed – practically all of the suns of both of those galaxies
> became possessed of planets.
>
> Edward E. Smith PhD, *Triplanetary*

TRIPLANETARY IS THE FIRST in Edward E. Smith's celebrated 'Lensman'
series of science fiction novels, and its opening paragraph reflects a
theory about the origin of planetary systems that was in vogue when
the book appeared in 1948. Even today it would be a powerful way to
start a science fiction novel; back then, it was breathtaking. The novels
themselves are early examples of 'widescreen baroque' space opera,
a cosmic battle between the forces of good (represented by Arisia)
and evil (Eddore) that takes six books to complete. The characters are
cardboard, the plots trite, but the action is enthralling, and at the time
the scope was unparalleled.

Today we no longer think that galactic collisions are needed to
create planets, though astronomers do see them as one of the four main
ways to make stars. The current theory of the formation of our own
solar system, and many other planetary systems, is different, yet no less
breathtaking than that opening paragraph. It goes like this.

Four and a half billion[1] years ago, a cloud of hydrogen gas six
hundred trillion kilometres across began to tear itself slowly to pieces.
Each piece condensed to create a star. One such piece, the solar nebula,
formed the Sun, together with its solar system of eight planets, five (so

far) dwarf planets, and thousands of asteroids and comets. The third rock from the Sun is our homeworld: Earth.

Unlike the fiction, it might even be true. Let's examine the evidence.

✦

The idea that the Sun and planets all condensed from a vast cloud of gas appeared remarkably early, and for a long time it was the prevailing scientific theory of their origins. When problems emerged, it went out of favour for nearly 250 years, but it has now been revived, thanks to new ideas and new data.

René Descartes is mainly famous for his philosophy – 'I think, therefore I am' – and his mathematics, notably coordinate geometry, which translates geometry into algebra and vice versa. But in his day 'philosophy' referred to many areas of intellectual activity, including science, which was *natural* philosophy. In his 1664[2] *Le Monde* (The World) Descartes tackled the origin of the solar system. He argued that initially the universe was a formless jumble of particles, circulating like whirlpools in water. One unusually large vortex swirled ever more tightly, contracting to form the Sun, and smaller vortices around it made the planets.

At a stroke, this theory explained two basic facts: why the solar system contains many separate bodies, and why the planets all go round the Sun in the same direction. Descartes's vortex theory doesn't agree with what we now know about gravity, but Newton's law wouldn't appear for another two decades. Emanuel Swedenborg replaced Descartes's swirling vortices with a huge cloud of gas and dust in 1734. In 1755 the philosopher Immanuel Kant gave the idea his blessing; the mathematician Pierre-Simon de Laplace stated it independently in 1796.

All theories of the origin of the solar system must explain two key observations. An obvious one is that matter has clumped together into discrete bodies: Sun, planets, and so on. A subtler one concerns a quantity known as *angular momentum*. This emerged from mathematical investigations into the deep implications of Newton's laws of motion.

The related concept of momentum is easier to understand. It governs the tendency of a body to travel at a fixed speed in a straight line when no forces are acting, as Newton's first law of motion states. Sports commentators use the term metaphorically: 'She's got the momentum now.' Statistical analysis offers remarkably little support for the proposal that a series of good scores tends to lead to more; commentators explain away the failures of their metaphor by observing (after the event) that the momentum has been lost again. In mechanics, the mathematics of moving bodies and systems, momentum has a very specific meaning, and one consequence is that you *can't* lose it. All you can do it transfer it to something else.

Think of a moving ball. Its speed tells us how fast it's moving: 80 kilometres per hour, say. Mechanics focuses on a more important quantity, velocity, which measures not just how fast it's going, but in which direction. If a perfectly elastic ball bounces off a wall, its speed remains unchanged but its velocity reverses direction. Its momentum is its mass multiplied by its velocity, so momentum also has both a size and a direction. If a light body and a massive one both move at the same speed in the same direction, the massive one has more momentum. Physically, it's then necessary to apply more force to change how the body is moving. You can easily bat away a ping-pong ball passing at 50 kph, but no one in their right mind would try that with a truck.

Mathematicians and physicists like momentum because, unlike velocity, it's conserved as a system changes over time. That is, the total momentum of the system remains fixed at whatever size, and direction, it was to start with.

That may sound implausible. If a ball hits a wall and bounces, its momentum changes direction, so it's not conserved. But the wall, much more massive, bounces a tiny bit too, and it bounces *the other way*. After that, other factors come into play, such as the rest of the wall, and I've kept up my sleeve the get-out clause: the conservation law only works when there are no external forces, that is, outside interference. This is how a body can acquire momentum to begin with: something gives it a shove.

Angular momentum is similar, but it applies to bodies that are rotating rather than moving in a straight line. Even for a single particle, its definition is tricky, but like momentum it depends both on the mass of the particle and its velocity. The main new feature is that it also depends on the axis of rotation – the line about which the particle is considered to be rotating. Imagine a spinning top. It spins around the line that runs through the middle of the top, so every particle of matter in the top rotates around this axis. The particle's angular momentum about that axis is its rate of spin multiplied by its mass. But the direction in which the angular momentum points is *along the spin axis*. That is, at right angles to the plane in which the particle is rotating. The angular momentum of the entire top, again considered about its axis, is obtained by adding together all of the angular momenta of its constituent particles, taking direction into account when necessary.

The size of a spinning system's total angular momentum tells us how strongly it's spinning, and the direction of the angular momentum tells us which axis it's spinning about, on average.[3] Angular momentum is conserved in any system of bodies subject to no external twisting forces (jargon: torque).

✦

This useful little fact has immediate implications for the collapse of a gas cloud: some good, some bad.

The good one is that, after some initial confusion, the gas molecules tend to spin in a single plane. Initially, each molecule has a certain amount of angular momentum about the centre of gravity of the cloud. Unlike a top, a gas cloud is not rigid, so these speeds and directions probably vary wildly. It's unlikely that all these quantities will cancel out perfectly, so initially the total angular momentum of the cloud is non-zero. The total angular momentum therefore points in some definite direction, and has a definite size. Conservation tells us that as the gas cloud evolves under gravity, its total angular momentum *doesn't change*. So the direction of the axis stays fixed, frozen in at the moment the cloud first formed. And the size – the total amount of spin, so to speak – is also frozen in. What can change is the distribution of the gas molecules. Every molecule of gas exerts a gravitational attraction on every other molecule, and the initially chaotic globular gas cloud collapses to form a flat disc, spinning about the axis like a plate on a pole in a circus.

This is good news for the solar nebula theory, because all the planets of the solar system have orbits that lie very close to the same plane – the ecliptic – and they all revolve in the same direction. That's why early astronomers guessed that the Sun and planets all condensed from a cloud of gas, after it had collapsed to create a protoplanetary disc.

Unfortunately for this 'nebular hypothesis', there is also some bad news: 99% of the solar system's angular momentum resides in the planets, with only 1% in the Sun. Although the Sun contains virtually the entire mass of the solar system, it's spinning quite slowly and its particles are relatively close to the central axis. The planets, though lighter, are much further away and move much faster, so they hog nearly all of the angular momentum.

However, detailed theoretical calculations show that a collapsing gas cloud doesn't do that. The Sun gobbles up most of the matter in the entire gas cloud, including a lot that was originally much further from the centre. So you'd expect it to have gobbled up the lion's share of the angular momentum ... which it spectacularly failed to do. Nevertheless, the current allocation of angular momentum, in which the planets get the lion's share, is entirely consistent with the dynamics of the solar system. It *works*, and it has done so for billions of years. There's no logical problem with the dynamics as such: just with how it all got started.

✦

One potential way out of this dilemma quickly emerged. Suppose the Sun formed *first*. Then it did gobble up pretty much all of the angular momentum in the gas cloud, because it gobbled up pretty much all of the gas. Afterwards it could acquire planets by *capturing* lumps of matter that passed nearby. If they were far enough away from the Sun, and moving at the right speed to be captured, the 99% figure would work out, just as it does today.

The main problem with this scenario is that it's very tricky to capture a planet. Any would-be planet that comes close enough will speed up as it approaches the Sun. If it manages not to fall into the Sun, it will swing round the Sun and be flung back out again. Since it's hard to capture just one planet, what price eight?

Perhaps, Count Buffon mused in 1749, a comet crashed into the Sun and splashed off enough material to create the planets. No, said Laplace in 1796: planets formed like that would eventually fall back into the Sun. The reasoning is similar to the 'no capture' line of argument, but in reverse. Capture is tricky because what comes down must go up again (unless it hits the Sun and gets engulfed). Splashing off is tricky because what goes up must come down. In any case we now know (as they didn't then) that comets are too lightweight to make a planet-sized splash, and the Sun is made of the wrong stuff.

In 1917 James Jeans suggested the tidal theory: a wandering star passed near the Sun and sucked out some of its material in a long, thin cigar. Then the cigar, which was unstable, broke into clumps that became the planets. Again, the Sun has the wrong composition; moreover this proposal requires a remarkably unlikely near collision, and doesn't endow the outermost planets with enough angular momentum to stop them falling back. Dozens of theories – all different, but all variations on similar themes – were proposed. Each fits some facts, and struggles to explain others.

By 1978, the apparently discredited nebular model was back in vogue. Andrew Prentice came up with a plausible solution to the angular momentum problem – recall: the Sun has way too little, the planets have too much. What's needed is a way to prevent angular momentum being conserved: to gain some or lose some. Prentice suggested that grains of dust concentrate near the centre of the disc of gas, and friction with those grains slows the rotation of the newly condensed Sun. Victor Safronov developed similar ideas around the same time, and his book on the topic led to the widespread adoption of the 'collapsing disc' model, in which the Sun and planets (and much else) all condensed out of a single massive gas cloud, pulled apart into clumps of many different sizes by its own gravity, modified by friction.

This theory has the merit of explaining why the inner planets (Mercury, Venus, Earth, Mars) are mainly rocky, while the outer ones (Jupiter, Saturn, Uranus, Neptune) are gas and ice giants. The lighter elements in the protoplanetary disc would accumulate further out than the heavy ones, though with much turbulent mixing. The prevailing theory for the giants is that a rocky core formed first, and its gravity attracted hydrogen, helium, and some water vapour, plus relatively small amounts of other material. However, models of planet formation have struggled to reproduce this behaviour.

In 2015 Harold Levison, Katherine Kretke, and Martin Duncan performed computer simulations supporting an alternative option: the cores slowly accreted from 'pebbles', lumps of rocky matter up to a metre across.[4] In theory this process can build up a core ten times the mass of the Earth in a few thousand years. Previous simulations had thrown up a different problem with this idea: it generates hundreds of planets the size of the Earth. The team showed that this problem is avoided if the pebbles come into being slowly enough to interact with each other gravitationally. Then the bigger ones scatter the rest out of the disc. Simulations with different parameters often produced between one and four gas giants at a distance of 5–15 AU from the Sun, consistent with the present structure of the solar system. One astronomical unit (AU) is the distance from the Earth to the Sun, and this is often a convenient way to grasp relatively small cosmic distances.

A good way to test the nebular model is to find out whether similar processes are going on elsewhere in the cosmos. In 2014 astronomers captured a remarkable image of the young star HL Tauri, 450 light years away in the constellation Taurus. The star is surrounded by concentric bright rings of gas, with dark rings in between. The dark rings are almost certainly caused by nascent planets sweeping up dust and gas. It would be difficult to find a more dramatic confirmation of a theory.

Atacama Large Millimeter Array image of HL Tauri, showing concentric rings of dust and gaps between them.

✦

It's easy to believe that gravity can cause things to clump, but how can it also pull them apart? Let's develop some intuition. Again, some serious mathematics, which we won't do here, confirms the general gist. I'll start with clumping.

A body of gas whose molecules attract each other *gravitationally* is very different from our usual experience of gases. If you fill a room with gas, it will very rapidly disperse so that the gas has the same density everywhere. You don't find odd pockets in your dining room where there's no air. The reason is that the molecules of air bounce around at random, and would quickly occupy any spare space. This behaviour is enshrined in the famous second law of thermodynamics, whose usual interpretation is that a gas is as disordered as possible. 'Disordered', in this context, has the connotation that everything should be thoroughly mixed together; that means that no region should be denser than any other.

To my mind this concept, technically known as entropy, is far too slippery to be captured by a simple word like 'disorder' – if only because 'evenly mixed' sounds to me like an *ordered* state. But for the moment, I'm going to toe the orthodox line. The mathematical formulation doesn't actually mention order or disorder at all, but it's too technical to discuss right now.

What holds in a room surely holds in a large room, so why not in a room the size of the entire universe? Indeed, in the universe itself? Surely the second law of thermodynamics implies that all of the gas in the universe ought to spread itself out uniformly into some sort of thin fog?

If that were true, it would be bad news for the human race, because we're not made from thin fog. We're distinctly lumpy, and we live on a rather bigger lump that orbits a lump so big that it sustains energetic nuclear reactions, producing heat and light. Indeed, people who dislike the usual scientific descriptions of the origin of humanity often invoke the second law of thermodynamics to 'prove' that we wouldn't be able to exist unless some hyper-intelligent being had deliberately manufactured us and arranged the universe to suit our requirements.

However, the thermodynamic model of gas in a room is not appropriate for working out how the solar nebula, or the entire universe, should behave. It has the wrong kind of interactions. Thermodynamics assumes that molecules notice each other only when they collide; then they bounce off each other. The bounces are perfectly elastic, meaning that no energy is lost, so the molecules continue bouncing forever. Technically, the forces that govern the interactions of molecules in a thermodynamic model of a gas are short range and repulsive.

Imagine a party where everyone is blindfolded and has their ears plugged, so that the only way to discover anyone else is present is to bump into them. Imagine, moreover, that everyone is hugely antisocial, so that when they do encounter someone else, both people immediately push each other away. It's plausible that after some initial bumping and wobbling they spread themselves around fairly evenly. Not all the time, because sometimes they come close together by accident, or even collide, but on average they stay spread out. A thermodynamic gas is like that, with absolutely gigantic numbers of molecules acting as people.

A gas cloud in space is more complicated. The molecules still bounce if they hit, but there's a second type of force: gravity. Gravity is ignored in thermodynamics because in that context its effects are negligibly small. But in cosmology, gravity is the dominant player, because there's an awful lot of gas. Thermodynamics keeps it gaseous, but gravity determines what the gas does on larger scales. Gravity is long range and attractive, almost the exact opposite of elastic bouncing. 'Long range' means that bodies interact even when they're far apart. The gravity of the Moon (and to a lesser extent the Sun) raises tides in the Earth's oceans, and the Moon is 400,000 kilometres away. 'Attractive' is straightforward: it causes the interacting bodies to move towards each other.

This is like a party where everyone can see everyone else from across the room – though less clearly from a distance – and as soon as they see anybody, they rush towards them. It's hardly surprising that a mass of gas interacting under gravity naturally becomes clumpy. In very small regions of the clumps, the thermodynamic model dominates, but on a larger scale the tendency to cluster dominates the dynamics.

If we're trying to work out what would happen to a hypothetical solar nebula, on the scale of solar systems or planets, we have to think about the long-range attractive force of gravity. The short-range repulsion between molecules that collide might tell us something about the state of a small region in a planet's atmosphere, but it won't tell us about the *planet*. In fact, it will mislead us into imagining that the planet should never have formed.[5]

Clumpiness is an inevitable consequence of gravity. A uniform spread is not.

✦

Since gravity causes matter to clump together, how can it also rip a molecular cloud apart? It seems contradictory.

The answer is that competing clumps can form at the same time. The mathematical arguments that support the collapse of a gas cloud into a flat, spinning disc assume we start with a region of gas that is roughly spherical – maybe shaped like an American football, but not like a dumbbell. However, a large region of gas will have occasional, randomly located places where by chance the matter is a little denser than elsewhere. Each such region acts as a centre, attracting more matter from its surroundings and exerting an ever-stronger gravitational force. The resulting cluster starts out fairly spherical and then collapses to a spinning disc.

However, in a large enough region of gas, several such centres can form. Although gravity is long range, its force drops off as the distance between bodies increases. So molecules are attracted to the nearest centre. Each centre is surrounded by a region in which its gravitational pull dominates. If there are *two* very popular people at the party, in opposite corners of the room, the party will split into two groups. So the gas cloud organises itself into a three-dimensional patchwork of attracting centres. These regions tear the cloud apart along their common boundaries. In practice what happens is a bit more complicated, and fast-moving molecules can escape the influence of the nearest centre and end up near a different one, but broadly speaking this is what we expect. Each centre condenses to form a star, and some of the debris surrounding it may form planets and other smaller bodies.

This is why an initially uniform cloud of gas condenses into a whole series of separate, relatively isolated star systems. Each system corresponds to one of the dense centres. But even then it's not totally straightforward. If two stars are close enough together, or approach each other for chance reasons, they can end up orbiting their common centre of mass. Now they form a binary star. In fact, systems of three or more stars can arise, loosely bound together by their mutual gravitation.

These multiple star systems, especially binary ones, are very common in the universe. The nearest star to the Sun, Proxima Centauri, is quite close (in astronomical terms) to a binary star called Alpha Centauri, whose individual stars are Alpha Centauri A and B. It seems likely that Proxima orbits both of these, but it probably takes half a million years to go once round its orbit. The distance between A and B is comparable to the distance from Jupiter to the Sun; it varies between about 11 and 36 AU.

In contrast, the distance from Proxima to either A or B is more like 15,000 AU, roughly a thousand times greater. Therefore, by inverse square law gravity, the force that A and B exert on Proxima is about one millionth of the force they exert on each other. Whether that's strong enough to keep Proxima in a stable orbit depends sensitively on what else is close enough to prise it from A and B's tenuous grasp. At any rate, we won't be around to see what happens.

✦

The early history of the solar system must have included periods of violent activity. The evidence is the huge number of craters on most bodies, especially the Moon, Mercury, Mars, and various satellites, showing that they were bombarded by innumerable smaller bodies. The relative ages of the resulting craters can be estimated statistically, because younger craters partially destroy older ones when they overlap, and most of the observed craters on these worlds are very ancient indeed. Even so, occasional new ones form, but most of these are very small.

The big problem here is to sort out the sequence of events that shaped the solar system. In the 1980s the invention of powerful computers and efficient and accurate computational methods permitted detailed mathematical modelling of collapsing clouds. Some sophistication is required because crude numerical methods fail to respect physical constraints such as conservation of energy. If this mathematical artefact causes energy to decrease slowly, the effect is like friction. Instead of following a closed orbit, a planet will spiral slowly into the Sun. Other quantities such as angular momentum must also be conserved. Methods that avoid this danger are of recent vintage. The most accurate ones are known as symplectic integrators, after a technical way to reformulate the equations of mechanics, and they conserve all relevant physical quantities *exactly*. Careful, precise simulations reveal plausible and very dramatic mechanisms for the formation of planets, which fit observations well. According to these ideas, the early solar system was very different from the sedate one we see today.

Astronomers used to think that once the solar system came into being, it was very stable. The planets trundled ponderously along preordained orbits and nothing much changed; the elderly system we see now is pretty much what it was in its youth. No longer! It's now thought that the larger worlds, the gas giants Jupiter and Saturn and the ice giants Uranus and Neptune, first appeared outside the 'frost line' where water freezes, but subsequently reorganised each other in a lengthy gravitational tug-of-war. This affected all of the other bodies, often in dramatic ways.

Mathematical models, plus a variety of other evidence from nuclear physics, astrophysics, chemistry, and many other branches of science, have led to the current picture: the planets didn't form as single clumps, but by a chaotic process of accretion. For the first 100,000 years, slowly growing 'planetesimals' swept up gas and dust, and created circular rings in the nebula by clearing out gaps between them. Each gap was littered with millions of these tiny bodies. At that point the planetesimals ran out of new matter to sweep up, but there were so many of them that they kept bumping into each other. Some broke up, but others merged; the mergers won and planets built up, piece by tiny piece.

In this early solar system, the giants were closer together than they are today, and millions of tiny planetesimals roamed the outer regions. Today the order of the giants, outwards from the Sun, is Jupiter, Saturn, Uranus, Neptune. But in one likely scenario it was originally Jupiter, Neptune, Uranus, Saturn. When the solar system was about 600 million years old, this cosy arrangement came to an end. All of the planets' orbital periods were slowly changing, and Jupiter and Saturn wandered into a 2:1 resonance – Jupiter's 'year' became exactly half that of Saturn. In general, resonances occur when two orbital or rotational periods are related by a simple fraction, here one half.[6] Resonances have a strong effect on celestial dynamics, because bodies in resonant orbits repeatedly align in exactly the same way, and I'll be saying a lot more about them later. This prevents disturbances 'averaging out' over long periods of time. This particular resonance pushed Neptune and Uranus outwards, and Neptune overtook Uranus.

This rearrangement of the larger bodies of the solar system disturbed the planetesimals, making them fall towards the Sun. All hell broke loose as planetesimals played celestial pinball among the planets. The giant planets moved out, and the planetesimals moved in. Eventually the planetesimals took on Jupiter, whose huge mass was decisive. Some planetesimals were flung out of the solar system altogether, while the rest went into long, thin orbits going out to huge distances. After that, everything mostly settled down, but the Moon, Mercury, and Mars still have battle scars resulting from the chaos.[7] And bodies of all shapes, sizes, and compositions were scattered far and wide.

Mostly settled down. It hasn't stopped. In 2008 Konstantin Batygin and Gregory Laughlin simulated the future of the solar system for 20 billion years, and the initial results revealed no serious instabilities.[8] Refining the numerical method to seek out potential instabilities, changing the orbit of at least one planet in a major way, they discovered a possible future in which Mercury hits the Sun about 1·26 billion years from now, and another in which Mercury's erratic movements eject Mars from the solar system 822 million years from now, followed by a collision between Mercury and Venus 40 million years later. The Earth sails serenely on, unaffected by the drama.

Early simulations mainly used averaged equations, not suitable for collisions, and ignored relativistic effects. In 2009 Jacques Laskar and Mickael Gastineau simulated the next 5 billion years of the solar system, using a method that avoided these problems,[9] but the results were much the same. Because tiny differences in initial conditions can have a large effect on long-term dynamics, they simulated 2,500 orbits, all starting within observational error of current conditions. In about 25 cases, the near resonance pumps up Mercury's eccentricity, leading either to a collision with the Sun, a collision with Venus, or a close encounter that radically changes the orbits of both Venus and Mercury. In one case, Mercury's orbit subsequently becomes less eccentric, causing it to destabilise all four of the inner planets within the next 3·3 billion years. The Earth is then likely to collide with Mercury, Venus, or Mars. And again there's a slight chance that Mars will be ejected from the solar system altogether.[10]

Inconstant Moon

It is the very error of the moon: She comes more nearer earth than she was wont, and makes men mad.

William Shakespeare, *Othello*

OUR MOON IS UNUSUALLY large.

It has a diameter just over one quarter of the Earth's, much larger than most other moons: so large, in fact, that the Earth–Moon system is sometimes referred to as a double planet. (Some jargon: Earth is the primary, the Moon is a satellite. Going up a level, the Sun is the primary of the planets in the solar system.) Mercury and Venus have no moons, while Mars, the planet most closely resembling Earth, has two tiny moons. Jupiter, the largest planet in the solar system, has 67 known moons, but 51 of them are less than 10 kilometres across. Even the largest, Ganymede, is less than one-thirtieth the size of Jupiter. Saturn is the most prolific in the satellite stakes, with over 150 moons and moonlets and a giant and complex ring system. But its largest moon, Titan, is only one twentieth as large as its primary. Uranus has 27 known moons, the largest being Titania, less than 1600 kilometres across. Neptune's only large moon is Triton, about one-twentieth of the planet's size; in addition astronomers have found 13 very small moons. Among the worlds of the solar system, only Pluto does better than us: four of its satellites are tiny, but the fifth, Charon, is about half the size of its primary.

The Earth–Moon system is unusual in another respect: its unusually large angular momentum. Dynamically, it has more 'spin' than it ought to. There are other surprises about the Moon, too, and we'll come to those in due course. The Moon's exceptional nature adds weight to a natural question: how did the Earth acquire its satellite?

The theory that fits the current evidence best is dramatic: the giant impact hypothesis. Early in its formation, our home planet was about 10% smaller than it is now, until a body about the size of Mars smashed into it, splashing off huge amounts of matter – initially much of it molten rock, in globules of all sizes, many of which merged as the rock began to cool. Part of the impactor united with the Earth, which became larger. Part of it became the Moon. The rest was dispersed elsewhere in the solar system.

Mathematical simulations support the giant impact scenario, while other theories fare less well. But in recent years, the giant impact hypothesis has started to run into trouble, at least in its original version. The origin of the Moon may still be up for grabs.

✦

The simplest theory is that the Moon accreted from the solar nebula, along with everything else, during the formation of the solar system. There was a lot of debris, in a huge range of sizes. As it started to settle down, larger masses grew by attracting smaller ones that merged with them after collisions. The planets formed this way, asteroids formed this way, comets formed this way, and moons formed this way. So presumably our Moon formed this way.

If so, however, it didn't form anywhere near its present orbit. The killer is angular momentum: there's too much of it. Another problem is the Moon's composition. As the solar nebula condensed, different elements were abundant at different distances. The heavier stuff stayed near the Sun, while radiation blew the lighter elements further out. That's why the inner planets are rocky, with iron–nickel cores, but the outer ones are mainly gas and ice – which is gas that got so cold, it froze. If Earth and Moon formed at roughly the same distance from the Sun, and at roughly the same time, they ought to have similar rocks in similar proportions. But the Moon's iron core is far smaller than the Earth's. In fact, Earth's total proportion of iron is eight times as large as the Moon's.

In the 1800s Charles Darwin's son George came up with another theory: in its early days the Earth, still molten, was spinning so fast that part of it broke off under the action of centrifugal force. He did the sums using Newtonian mechanics, and predicted that the Moon must be moving away from the Earth, which turns out to be true. This event would have left a large scar, and there was an obvious candidate: the Pacific Ocean. However, we now know that Moon rock is much older than the oceanic crustal material in the Pacific. That rules out the Pacific basin, but not necessarily Darwin's fission theory.

Plenty of other scenarios have been suggested, some rather wild. Perhaps a natural nuclear reactor (at least one is known to have existed, by the way[1]) went critical, exploded, and expelled the lunar material. If the reactor was near the boundary between mantle and core, close to the equator, a lot of Earth rock would have gone into equatorial orbit. Or perhaps Earth originally had two moons, which collided. Or we stole a moon from Venus, which neatly explains why Venus doesn't have one. Though it fails to explain why, if that theory is correct, Earth originally didn't.

A less dramatic alternative is that the Earth and the Moon formed separately, but later the Moon came close enough to the Earth to be captured by its gravity. This idea has several things going for it. The Moon has the right size, and it's in a sensible orbit. Moreover, capture explains why the Moon and Earth are 'tidally locked' by their mutual gravity, so that the same side of the Moon always faces the Earth. It wobbles a bit (jargon: libration), but that's normal with tidal locking.

The main issue is that although gravitational capture sounds reasonable (bodies attract each other, after all) it's actually rather unusual. The motion of celestial bodies involves hardly any friction – there's some, for example with the solar wind, but its dynamic effects are minor – so energy is conserved. The (kinetic) energy that a 'falling' body acquires as it approaches another one, pulled by their mutual gravitational interaction, is therefore big enough for the body to *escape* that pull again. Typically the two bodies approach, swing round each other, and separate.

Alternatively, they collide.

Clearly the Earth and Moon did neither.

There are ways round this problem. Perhaps the early Earth had a huge extended atmosphere, which slowed the Moon down when it came close, without breaking it up. There's a precedent: Neptune's moon Triton is exceptional not just in its size, compared with the planet's other moons, but in its direction of motion, which is 'retro-grade' – the opposite way round to most of the bodies of the solar system, including all of the planets. Astronomers think that Triton was captured by Neptune. Originally, Triton was a Kuiper belt object (KBO), the name given to a swarm of smallish bodies orbiting beyond Neptune. This is an origin that it probably shares with Pluto. If so, captures do occur.

Another observation constrains the possibilities even more. Although the overall geological compositions of the Earth and Moon are very different, the detailed composition of the Moon's surface rocks is remarkably similar to that of the Earth's mantle. (The mantle lies between the continental crust and the iron core.) Elements have 'isotopes' that are chemically almost identical, but differ in the particles forming the atomic nucleus. The commonest oxygen isotope, oxygen-16, has eight protons and eight neutrons. Oxygen-17 has an extra neutron, and oxygen-18 has a second extra neutron. When rocks form, oxygen is incorporated through chemical reactions. Samples of Moon rock brought back by *Apollo* astronauts have the same ratios of oxygen and other isotopes as the mantle.

In 2012 Randall Paniello and coworkers analysed zinc isotopes in lunar material, finding that it has less zinc than the Earth, but a higher proportion of heavy isotopes of zinc. They concluded that the Moon has lost zinc by evaporation.[2] Again, in 2013 a team under Alberto Saal reported that atoms of hydrogen included in lunar volcanic glass and olivine have very similar isotope ratios to Earth's water. If the Earth and Moon originally formed separately, it would be unlikely for their isotope ratios to be so alike.

The simplest explanation is that these two bodies have a common origin, despite differences in their cores. However, there's an alternative: perhaps they had distinct origins, and their composition was different when they formed, but later they were mixed together.

✦

Let's review the evidence that needs explaining. The Earth–Moon system has unusually large angular momentum. The Earth has a lot more iron than the Moon, yet the lunar surface has very similar isotope ratios to the Earth's mantle. The Moon is unusually large, and tidally locked to its primary. Any viable theory has to explain, or at least be consistent with, these observations to be remotely plausible. And *none* of the simple theories does that. It's the Sherlock Holmes cliché: 'When you have eliminated the impossible, then whatever remains, however unlikely, must be the truth.' And the simplest explanation that does fit the evidence is something that, until the late twentieth century, astronomers would have rejected because it seemed so improbable. Namely, Earth collided with something else, so massive that the collision melted both bodies. Some of the molten rock splashed off to form the Moon, and what merged with the Earth contributed much of its mantle.

This giant impact hypothesis, in its currently favoured incarnation, dates from 1984. The impactor even has a name: Theia. However, unicorns have a name but don't exist. If Theia ever existed, the only remaining traces are on the Moon and deep in the Earth, so the evidence has to be indirect.

Few ideas are truly original, and this one goes back at least to Reginald Daly, who objected to Darwin's fission theory because when you do the sums properly, the Moon's current orbit doesn't trace all the way back to the Earth when you run time backwards. An impact, Daly proposed, would work a lot better. The main apparent problem, at that time, was: impact with what? In those days, astronomers and mathematicians thought that the planets formed in pretty much their present orbits. But as computers got more powerful, and the implications of Newton's mathematics could be explored in more realistic settings, it became clear that the early solar system kept changing dramatically. In 1975 William Hartmann and Donald Davis performed calculations suggesting that after the planets had formed, several smaller bodies were left over. These might be captured and become moons, or they might collide, either with each other or with a planet. Such a collision, they said, could have created the Moon, and is consistent with many of its known properties.

In 1976 Alastair Cameron and William Ward proposed that another planet, about the size of Mars, collided with the Earth, and some of the material that splashed off aggregated to form the Moon.[3] Different constituents would behave differently under the massive forces and heat generated by the impact. Silicate rock (on either body) would vaporise, but the Earth's iron core, and any metallic core that the impactor might possess, would not. So the Moon would end up with much less iron than the Earth, but the surface rocks of the Moon and the Earth's mantle, condensing back from the vaporised silicates, would be extremely similar in composition.

In the 1980s Cameron and various coworkers carried out computer simulations of the consequences of such an impact, showing that a Mars-sized impactor – Theia – fits observations best.[4] At first it seemed plausible that Theia could splash off a chunk of the Earth's mantle, while contributing very little of its own material to the rocks that became the Moon. That would explain the very similar composition of these two types of rock. Indeed, this was seen as strong confirmation of the giant impact hypothesis.

Until a few years ago, most astronomers accepted this idea. Theia smashed into the primeval Earth very soon (in cosmological terms) after the formation of the solar system, between 4·5 and 4·45 billion years ago. The two worlds didn't collide head-on, but at an angle of about 45 degrees. The collision was relatively slow (again in cosmo-logical terms): around 4 kilometres per second. Calculations show that if Theia had an iron core, this would have merged with the main body of the Earth, and being denser than the mantle, it would have sunk and coalesced with the Earth's core; remember, the rocks were all molten at this stage. That explains why Earth has a lot more iron than the Moon. About one fifth of Theia's mantle, and quite a lot of Earth's silicate rock, was thrown into space. Half of that ended up orbiting the Earth, and aggregated to create the Moon. The other half escaped from the Earth's gravity and orbited the Sun. Most of it stayed in orbits roughly similar to the Earth's, so it collided with the Earth or the newly formed Moon. Many of the lunar craters were created by these secondary impacts. On Earth, however, erosion and other processes erased most impact craters.

The impact gave the Earth extra mass, and a lot of extra angular momentum: so much that it spun once every five hours. The Earth's slightly oblate shape, squashed at the poles, exerted tidal forces that aligned the Moon's orbit with the Earth's equator, and stabilised it there.

Measurements show that the crust of the Moon on the side that is now turned away from Earth is thicker. The thinking is that some of the splashed-off material in Earth orbit initially failed to be absorbed into what became the Moon. Instead, a second, smaller moon collected together in a so-called 'Lagrange point', in the same orbit as the Moon but 60 degrees further along it (see Chapter 5). After 10 million years, as both bodies drifted slowly away from the Earth, this location became unstable and the smaller moon collided with the larger one. Its material spread across the far side of the Moon, thickening the crust.

✦

I've used the words 'simulation' and 'calculation' quite a lot, but you can't do a sum unless you know what you want to calculate, and you can't simulate something by just 'putting it on the computer'. Someone has to set up the calculation, in exquisite detail; someone has to write software that tells the computer how to do the sums. These tasks are seldom straightforward.

Simulating a cosmic impact is a horrendous computational problem. The matter involved can be solid, liquid, or vapour, and different physical rules apply to each case, requiring different mathematical formulations. At least four types of matter are involved: core and mantle for each of Theia and Earth. The rocks, in whatever state, can fragment or collide. Their motion is governed by 'free boundary conditions', meaning that the fluid dynamics does not take place in a fixed region of space with fixed walls. Instead, the fluid itself 'decides' where its boundary is, and its location changes as the fluid moves. Free boundaries are much harder to handle than fixed ones, both theoretically and computationally. Finally, the forces that act are gravitational, so they are nonlinear. That is, instead of changing proportionately to distance, they change according to the inverse square law. Nonlinear equations are notoriously harder than linear ones.

Traditional pencil-and-paper mathematical methods can't hope to solve even simplified versions of the problem. Instead, fast computers with lots of memory use numerical methods to approximate the problem, and then do a lot of brute force calculations to get an approximate answer. Most simulations model the colliding bodies as droplets of sticky fluid, which can break up into smaller droplets or merge to create larger ones. The initial drops are planet-sized; the droplets are smaller, but only in comparison to planets. They're still fairly big.

A standard model for fluid dynamics goes back to Leonhard Euler and Daniel Bernoulli in the 1700s. It formulates the physical laws of fluid flow as a partial differential equation, which describes how the velocity of the fluid at each point in space changes over time in response to the forces that act. Except in very simple cases, it's not possible to find formulas that solve the equation, but very accurate computational methods have been devised. A major issue is the nature of the model, which in principle requires us to study the fluid velocity at every point in some region of space. However, computers can't do infinitely many calculations, so we 'discretise' the equation: approximate it by a related equation involving only a finite number of points. The simplest method uses the points of a grid as a representative sample of the entire fluid, and keeps track of how the velocity changes at the grid points. This approximation is good if the grid is fine enough.

Unfortunately this approach isn't great for colliding drops, because the velocity field becomes discontinuous when the drops break up. A cunning variant of the grid method comes to the rescue. It works even when droplets fragment or merge. This method, called smoothed particle hydrodynamics, breaks the fluid up into neighbouring 'particles' – tiny regions. But instead of using a fixed grid, we follow the particles as they respond to the forces that act. If nearby particles move with much the same speed and direction, they're in the same droplet and are going to stay in that droplet. But if neighbouring particles head off in radically different directions, or have significantly different velocities, the droplet is breaking up.

The mathematics makes this work by 'smoothing' each particle into a sort of soft fuzzy ball (jargon: spherical overlapping kernel function) and superposing these balls. The motion of the fluid is calculated by combining the motions of the fuzzy balls. Each ball can be represented

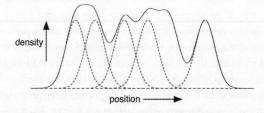

Representing the density of a fluid (solid line) as a sum of small fuzzy droplets (dashed bell-shaped curves).

by its central point, and we have to calculate how these points move as time passes. Mathematicians call this kind of equation an n-body problem, where n is the number of points, or, equivalently, the number of fuzzy balls.

✦

All very well, but n-body problems are hard. Kepler studied a two-body problem, the orbit of Mars, and deduced it's an ellipse. Newton proved mathematically that when two bodies move under inverse square law gravity, both of them orbit in ellipses about their common centre of mass. But when mathematicians of the eighteenth and nineteenth centuries tried to understand the three-body problem – Sun, Earth, Moon is the basic case – they discovered it's nowhere near as neat and tidy. Even Delaunay's mammoth formula is only an approximation. In fact, the orbits are typically chaotic – highly irregular – and there are no nice formulas or classical geometric curves to describe them. See Chapter 9 for more about chaos.

To model a planetary collision realistically, the number n of fuzzy balls must be large – a thousand, or better still a million. Computers can calculate with big numbers, but here n doesn't characterise the numbers that appear in the sums: it measures how *complicated* the sums are. Now we run into the 'curse of dimensionality', where the dimension of a system is how many numbers you need to describe it.

Suppose we use a million balls. It takes six numbers to determine the state of each ball: three for its coordinates in space, three more for the components of its velocity. That's 6 million numbers, just to define the state at any instant. We want to apply the laws of mechanics and gravity to predict the future motion. These laws are differential equations, which determine the state a tiny period of time into the future, given the state now. Provided the time step into the future is very small, a second perhaps, the result will be very close to the correct future state. So now we're doing a sum with 6 million numbers. More precisely, we're doing 6 *million sums* with 6 million numbers: one sum for each number required for the future state. So the complexity of the calculation is 6 million *multiplied by* 6 million. That's 36 trillion. And this calculation tells us only what the next state is, one second into the future. Do the same again and we find out what happens two seconds from now, and so on. To find out what happens a thousand years ahead, we're looking at a period of about 30 billion seconds, and the complexity of the calculation is 30 billion times 36 trillion – about 10^{24}, one septillion.

And that's not the worst of it. Although each individual step may be a good approximation, there are now so many steps that even the tiniest error might grow, and big calculations take a lot of time. If the computer could do one step per second – that is, work in 'real time' – the sums would take a thousand years. Only a supercomputer could even come close to that. The only way out is to find a cleverer way to do the sums. In the early stages of the impact, a time step as short as a second might be needed because everything is a complicated mess. Later, a longer time step might be acceptable. Moreover, once two points get far enough apart, the force between them is so small that it might be possible to neglect it altogether. Finally – and this is where the main improvement comes from – the entire calculation might be simplified by setting it up in a more cunning way.

The earliest simulations made a further simplification. Instead of doing the calculations for three-dimensional space, they cut the problem down to two dimensions by assuming everything happens in the plane of the Earth's orbit. Now two circular bodies collide instead of two spherical ones. This simplification offers two advantages. That 6 million becomes only 4 million (four numbers per fuzzy ball). Better still, you no longer need a million balls; maybe 10,000 are enough. Now you have 40,000 in place of 6 million, and the complexity reduces from 36 trillion to 1·6 billion.

Oh, and one more thing…

It's not enough to do the calculation once. We don't know the mass of the impactor, its speed, or the direction it's coming from when it hits. Each choice requires a new calculation. This was a particular limitation of early work, because computers were slower. Time on a supercomputer was expensive, too, so research grants allowed only a small number of runs. Consequently the investigator had to make some good guesses, right at the start, based on rule-of-thumb considerations such as 'can this assumption give the right size for the final angular momentum?' And then hope.

The pioneers overcame these obstacles. They found a scenario that worked. Later work refined it. The origin of the Moon had been solved.

✦

Or had it?

Simulating the giant impact theory of the Moon's formation involves two main phases: the collision itself, creating a disc of debris, and the subsequent accretion of part of this disc to form a compact lump, the nascent Moon. Until 1996 researchers confined their calculations to the first phase, and their main method was smoothed particle hydrodynamics. Robin Canup and Erik Asphaug, writing in 2001, stated[5] that this method 'is well suited to intensely deforming systems evolving within mostly empty space,' which is exactly what we want for this phase of the problem.

Because these simulations are big and difficult, investigators contented themselves with working out what happened immediately after the impact. The results depend on many factors: the mass and speed of the impactor, the angle at which it hits the Earth, and the rotational speed of the Earth, which, several billion years ago, might well have been different from what it is today. The practical limitations of n-body computations meant that, to begin with, many alternatives were not explored. To keep the computations within bounds, the first models were two-dimensional. Then it was a matter of seeking cases where the impactor kicked a lot of material from the Earth's mantle into space. The most convincing example involved a Mars-sized impactor, so this became the prime contender.

All of these giant impact simulations had one feature in common: the impact created a huge disc of debris orbiting the Earth. The simulations usually modelled the dynamics of this disc for only a few orbits, enough to show that plenty of debris stayed in orbit rather than crashing back down again or heading off into outer space. It was *assumed* that many of the particles in the debris disc would eventually aggregate to form a large body, and that body would become the Moon, but no one checked this assumption because tracking the particles further would be too expensive and time-consuming.

Some of the later work made the tacit assumption that the main parameters – mass of impactor and so on – had already been sorted out by this pioneering work, and concentrated on calculating extra detail, rather than looking at alternative parameters. The pioneering work became a kind of orthodoxy, and some of its assumptions ceased to be questioned. The first sign of trouble came early on. The only scenarios that gave a plausible fit to observations required the impactor to graze the Earth rather than smash into it head on, so the impactor could not have been in the Earth's orbital plane. The two-dimensional model is inadequate, and only a full three-dimensional simulation can do the job. Fortunately the powers of supercomputers evolve rapidly, and with enough time and expense it became possible to analyse collisions in three-dimensional models.

However, most of these improved simulations showed that the Moon should contain a lot of rock from the *impactor*, and much less from the Earth's mantle. So the original simple explanation of the similarity between Moon rock and the mantle became much less convincing: it seemed to require Theia's mantle to be amazingly similar to that of the Earth. Some astronomers nevertheless maintained that this was what must have happened, neatly forgetting that such a similarity between Earth and Moon was one of the puzzles that the theory was supposed to explain. If it didn't wash for the Moon, why was it acceptable for Theia?

There's a partial answer: maybe Theia and the Earth originally formed at about the same distance from the Sun. The objections raised earlier for the Moon don't apply. There's no issue with angular momentum, because we haven't a clue what the other chunks of Theia did after the impact. And it's reasonable to assume that bodies that formed at similar locations in the solar nebula have similar compositions. But it's still hard to explain why Earth and Theia stayed separate for long enough to become planets in their own right – but then collided. It's not impossible, but it doesn't look likely.

A different theory seems more plausible, because it makes no assumptions about Theia's composition. Suppose that after the silicate rocks vaporised, and before they started to aggregate, they were thoroughly mixed. Then both Earth and Moon would have received donations of very similar rock. Calculations indicate that this idea works only if the vapour stays around for about a century, forming a kind of shared atmosphere spread along the common orbit of Theia and the Earth. Mathematical studies are under way to decide whether this theory is dynamically feasible.

Be that as it may, the original idea that the impactor splashed off a chunk of Earth's mantle, but did not itself contribute much to the eventual Moon, would be much more convincing. So astronomers sought alternatives, still involving a collision, but based on very different assumptions. In 2012 Andreas Reufer and coworkers analysed the effects of a fast-moving impactor much larger than Mars that sideswipes the Earth instead of colliding head-on.[6] Very little of the material splashed off comes from the impactor, the angular momentum works out fine, and the composition of the mantle and the Moon are even more similar than previously thought. According to a new analysis of *Apollo* lunar rock by Junjun Zhang's team, the ratio of titanium-50 and titanium-47 isotopes is the same as for the Earth within four parts per million.[7]

Other possibilities have been studied as well. Matja Cuk and coworkers have shown that the correct chemistry of Moon rocks and total angular momentum could have arisen from a collision with a smaller impactor, provided the Earth was spinning much faster than it is today. The spin changes the amount of rock that splashes off, and which body it comes from. After the collision, gravitational forces from the Sun and Moon could have slowed the Earth's spin. On the other hand, Canup has found convincing simulations in which the Earth was spinning only marginally faster than today, by assuming the impactor was significantly bigger than Mars. Or perhaps two bodies, five times the size of Mars collided, then recollided, creating a large disc of debris that eventually formed the Earth and Moon. Or...

✦

Or possibly the original impactor theory is correct, Theia *did* have much the same composition as the Earth, and that wasn't a coincidence at all.

In 2004 Canup[8] showed that the most plausible type of Theia should have about one-sixth of the mass of the Earth, and that four-fifths of the resulting Moon's material should have come from Theia. This implies that Theia's chemical composition must have been as close to that of the Earth as the Moon's is. This seems very unlikely: the bodies of the solar system differ considerably from each other, so what was different about Theia? As we've seen, a possible answer is that Earth and Theia formed under similar conditions – at a similar distance from the Sun, so they both swept up the same stuff. Moreover, being in roughly the same orbit improves the chance of a collision.

On the other hand, could two large bodies form in the same orbit? Wouldn't one of them win by sweeping up most of the available material? You can argue about this forever … or you can do the sums. In 2105 Alessandra Mastrobuono-Battisti and coworkers used n-body methods to make 40 simulations of the late stages of planetary accretion.[9] By then, Jupiter and Saturn are fully formed, they've sucked up most of the gas and dust, and planetesimals and larger 'planetary embryos' are coming together to form the really large bodies. Each run started with about 85–90 planetary embryos and 1000–2000 planetesimals, lying in a disc between 0·5 and 4·5 AU. The orbits of Jupiter and Saturn were inclined slightly to this disc, and the inclinations differed between runs.

In most runs, about three or four rocky inner planets formed within 100–200 million years, as the embryos and planetesimals merged. The simulation kept track of each world's feeding zone, the region from which its components were gobbled up. On the assumption that the chemistry of the solar disc depends mainly on distance from the Sun, so that bodies in equidistant orbits have much the same composition, we can compare the chemical compositions of impacting bodies. The team focused on how each of the three or four surviving planets compares to its most recent impactor. Tracking back through these bodies' feeding zones leads to probability distributions for the composition of each body. Then statistical methods determine how similar these distributions are. The impactor and the planet have much the same composition in about one-sixth of the simulations. Taking into account the likelihood that some of the proto-planet also gets mixed into the Moon, this figure doubles to about one-third. In short: there is about *one chance in three* that Theia would have had the same chemistry as Earth. This is entirely plausible, so despite previous concerns, the similar chemistry of Earth's mantle and the Moon's surface rock is, in fact, consistent with the original giant impact scenario.

Right now, we have an embarrassment of riches: several distinct giant impact theories, all in good agreement with the main evidence. Which, if any, is correct remains to be seen. But to get both the chemistry and the angular momentum right, a large impactor seems unavoidable.

4

The Clockwork Cosmos

But should the Lord Architect have left that space empty? Not at all.

Johann Titius, in Charles Bonnet's *Contemplation de la Nature*

NEWTON'S *PRINCIPIA* ESTABLISHED the value of mathematics as a way to understand the cosmos. It led to the compelling notion of a clockwork universe, in which the Sun and planets were created in their present configuration. The planets went round and round the Sun in roughly circular orbits, nicely spaced so that they didn't run into each other – didn't even come close. Although everything jiggled about a bit, thanks to each planet's gravity tugging every other planet, nothing important changed. This view was encapsulated in a delightful gadget known as an orrery: a desktop machine in which tiny planets on sticks moved round and round the central Sun, driven by cogwheels. Nature was a gigantic orrery, with gravity for gears.

Mathematically minded astronomers knew that it wasn't quite that simple. The orbits weren't exact circles, they didn't even lie in the same plane, and some of the jiggles were quite substantial. In particular the two largest planets in the solar system, Jupiter and Saturn, were engaged in some kind of long-term gravitational tug-of-war, pulling each other first ahead of their usual positions in their orbits, then behind, over and over again. Laplace explained this around 1785. The two giants are close to a 5:2 resonance, in which Jupiter goes round the Sun five times while Saturn goes round twice. Measuring their positions in orbit as angles, the difference

$$2 \times \text{angle for Jupiter} - 5 \times \text{angle for Saturn}$$

is close to zero – but, as Laplace explained, it's not exactly zero. Instead, it slowly changes, completing a full circle every 900 years. This effect became known as the 'great inequality'.

Laplace proved that the interaction doesn't produce large changes to the eccentricity or inclination of either planet's orbit. This kind of result led to a general feeling that the current arrangement of the planets is stable. It would be much the same far into the future, and it had always been that way in the past.

Not so. The more we learn about the solar system, the less it looks like clockwork, and the more it looks like some bizarre structure that, although *mostly* well behaved, occasionally goes completely crazy. Remarkably, these weird gyrations don't cast doubt on Newton's law of gravity: they are *consequences* of it. The law itself is mathematically neat and tidy, simplicity itself. But what it leads to is not.

✦

To understand the origins of the solar system, we must explain how it arose and how its multifarious bodies are arranged. At first sight they're a pretty eclectic lot – each world is a one-off, and the differences outweigh the similarities. Mercury is a hot rock that revolves three times every two orbits, a 3:2 spin–orbit resonance. Venus is an acid hell whose entire surface reformed a few hundred million years ago. Earth has oceans, oxygen, and life. Mars is a frigid desert with craters and canyons. Jupiter is a giant ball of coloured gases making decorative stripes. Saturn is similar, though less dramatic, but in compensation it has gorgeous rings. Uranus is a docile ice giant and it spins the wrong way. Neptune is another ice giant, with encircling winds that exceed 2000 kilometres per hour.

However, there are also tantalising hints of order. The orbital distances of the six classical planets, in astronomical units, are:

Mercury	0·39
Venus	0·72
Earth	1·00
Mars	1·52
Jupiter	5·20
Saturn	9·54

The numbers are a bit irregular, and at first it's hard to find a pattern, even if it occurs to you to look. But in 1766 Johann Titius spotted something interesting in these numbers, and described it in his translation of Charles Bonnet's *Contemplation de la Nature*:

> Divide the distance from the Sun to Saturn into 100 parts; then Mercury is separated by four such parts from the Sun, Venus by 4+3 = 7 such parts, the Earth by 4+6 = 10, Mars by 4+12 = 16. But notice that from Mars to Jupiter there comes a deviation from this so exact progression. From Mars there follows a space of 4+24 = 28 such parts, but so far no planet was sighted there. ... Next to this for us still unexplored space there rises Jupiter's sphere of influence at 4+48 = 52 parts; and that of Saturn at 4+96 = 100 parts.

Johann Bode mentioned the same numerical pattern in 1772 in his *Anleitung zur Kenntniss des Gestirnten Himmels* (Manual for Knowing the Starry Sky), and in later editions he credited it to Titius. Despite that, it's often called Bode's law. A better term, now in general use, is Titius–Bode law.

This rule, which is purely empirical, relates planetary distances to a (nearly) geometric sequence. Its original form started with the sequence 0, 3, 6, 12, 24, 48, 96, 192, in which each number after the second is twice its predecessor, and added 4 to them all, getting: 4, 7, 10, 16, 28, 52, 100. However, it's useful to bring these numbers into line with current units of measurement (AU) by dividing them all by ten, giving: 0·4, 0·7, 1·0, 1·6, 2·8, 5·2, 10·0. These numbers fit the spacing of the planets surprisingly well, except for a gap corresponding to 2·8. Titius thought he knew what must be in that gap. The portion of his remark that I replaced by an ellipsis (...) reads:

> But should the Lord Architect have left that space empty? Not at all. Let us therefore assume that this space without doubt belongs to the still undiscovered satellites of Mars, let us also add that perhaps Jupiter still has around itself some smaller ones which have not been sighted yet by any telescope.

We now realise that the satellites of Mars will be found close to Mars, and ditto for Jupiter, so Titius was a bit shy of the mark in some respects, but the proposal that *some* body ought to occupy the gap was spot on. However, no one took it seriously until Uranus was discovered in 1781, and it also fitted the pattern. The predicted distance is 19·6; the actual one is 19·2.

Encouraged by this success, astronomers started looking for a previously unobserved planet circling the sun at about 2·8 times the radius of Earth's orbit. In 1801 Giuseppe Piazzi found one – ironically, just before a systematic search got under way. It was given the name Ceres, and we take up its story in Chapter 5. It was smaller than Mars, and much smaller than Jupiter, but it was *there*.

To make up for its diminutive stature, it was not alone. Three more bodies – Pallas, Juno, and Vesta – were soon found at similar distances. These were the first four asteroids, or minor planets, and they were soon followed by many more. About 200 of them are more than a kilometre across, over 150 million at least 100 metres across are known, and there are expected to be millions that are even smaller. They famously form the asteroid belt, a flat ring-shaped region between the orbits of Mars and Jupiter.

Other small bodies exist elsewhere in the solar system, but the first few discoveries added weight to Bode's view that the planets are distributed in a regular manner. The subsequent discovery of Neptune was motivated by discrepancies in the orbit of Uranus, not by the Titius–Bode law. But the law predicted a distance of 38·8, reasonably close to the actual distance, between 29·8 and 30·3. The fit is poorer but acceptable. Then came Pluto: theoretical distance 77·2, actual distance between 29·7 and 48·9. Finally, the Titius–Bode 'law' had broken down.

Other typical features of planetary orbits had also broken down. Pluto is very strange. Its orbit is highly eccentric and tilted a whopping 17 degrees away from the ecliptic. Sometimes Pluto even comes *inside* the orbit of Neptune. Unusual features like this recently led to Pluto being reclassified as a dwarf planet. In partial compensation, Ceres also became a dwarf planet, not a mere asteroid (or minor planet).

Despite its mix of success and failure, the Titius–Bode law poses an important question. Is there some mathematical rationale to the spacing of the planets? Or could they, in principle, have been spaced in any desired manner? Is the law a coincidence, a sign of an underlying pattern, or a bit of both?

✦

The first step is to reformulate the Titius–Bode law in a more general and slightly modified way. Its original form has an anomaly: the use of 0 as the first term. To get a geometric sequence this ought to be 1·5. Although this choice makes the distance for Mercury 0·55, which is less accurate, the whole game is empirical and approximate, so it makes more sense to keep the mathematics tidy and use 1·5. Now we can express the law in a simple formula: the distance from the Sun to the nth planet, in astronomical units, is

$$d = 0·075 \times 2^n + 0·4$$

Now we must do a few sums. In the grand scheme of things, 0·4 AU doesn't make much difference for the more distant planets, so we remove it to get $d = 0·075 \times 2^n$. This is an example of a power law formula, which in general looks like $d = ab^n$, where a and b are constants. Take logarithms:

$$\log d = \log a + n \log b$$

Using n and $\log d$ as coordinates, this is a straight line with slope $\log b$, meeting the vertical axis at $\log a$. So the way to spot a power law is to perform a 'log/log plot' of $\log d$ against n. If the result is close to a straight line, we're in business. Indeed, we can do this for quantities other than the distance d, for example the period of revolution around the star or the mass.

If we try this for the distances of planets, including Ceres and Pluto, we obtain the left-hand picture. Not far from a straight line, as we'd expect from the Titius–Bode law. What about the masses, shown in the right-hand picture? This time the log/log plot is very different. No sign of a straight line – or of any clear pattern.

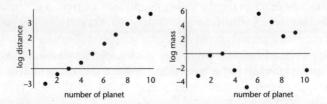

Left: Log/log plot of planetary distances lies close to a straight line.
Right: Log/log plot of planetary masses doesn't look like a straight line.

The orbital periods? A nice straight line again: see the left-hand picture. However, that's no surprise, because Kepler's third law relates to period to the distance in a manner that preserves power law relationships. Searching further afield, we examine the five main moons of Uranus, and get the right-hand picture. Power law again.

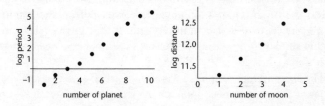

Left: Log/log plot of planetary periods lies close to a straight line.
Right: Log/log plot of distances of moons of Uranus lies close to a straight line.

✦

Coincidence, or something deeper? Astronomers are divided. At best there seems to be a *tendency* towards power law spacings. They're not universal.

There may be a rational explanation. The most likely one starts from the idea that the dynamics of a random system of planets depends crucially on resonances: cases where two planets have orbital periods in some simple fractional relationship. For example, one period might be 3/5 of the other, a 5:3 resonance.[1] Ignoring all other bodies, these two planets will keep lining up with each other, along the radial line from the star, at regular intervals, because five rotations of one match three of the other perfectly. Over long periods of time the resulting small disturbances will accumulate, so the planets will tend to change their orbits. For period ratios that are not simple fractions, on the other hand, the disturbances tend to cancel out, because there's no dominant direction for the force of gravity acting between the two worlds.

This isn't just a vague suggestion: detailed calculations and an extensive body of mathematical theory support it. To a first approximation, the orbit of a celestial body is an ellipse. At the next level of approximation, the ellipse precesses: its major axis slowly rotates.

Approximating even more accurately, the dominant terms in formulas for the motion of celestial bodies come from secular resonances – more general types of resonant relation among the periods with which the orbits of several bodies precess.

Precisely how resonant bodies move depends on the ratio of their periods, as well as their locations and velocities, but often the result is to clear out such orbits. Computer simulations indicate that randomly spaced planets tend to evolve into positions that satisfy relationships roughly similar to the Titius–Bode law, as resonances sweep out gaps. But it's all a bit vague.

The solar system contains several 'miniature' systems, namely, the moons of the giant planets. Jupiter's three largest satellites, Io, Europa, and Ganymede, have orbital periods in ratios close to 1:2:4, each twice the previous one (see Chapter 7). The fourth, Callisto, has a period slightly less than twice that of Ganymede. By Kepler's third law, the orbital radii are similarly related, except that the multiplier 2 has to be replaced by its 2/3 power, which is 1·58. That is, the orbital radius of each satellite is roughly 1·58 times that of the previous one. This is a case where resonance stabilises orbits instead of clearing them out, and the ratio of distances is 1·58 rather than the 2 of the Titius–Bode law. But the spacings still satisfy a power law. The same goes for the moons of Saturn and Uranus, as Stanley Dermott pointed out in the 1960s.[2] Such a spacing is called 'Dermott's law'.

Power law spacings are a more general pattern that includes a good approximation to the Titius–Bode law. In 1994 Bérengère Dubrulle and François Graner derived power law spacings for typical collapsing solar nebulas[3] by applying two general principles. Both depend on symmetry. The cloud is axially symmetric, and the matter distribution is much the same on all scales of measurement, a symmetry of scale. Axial symmetry makes dynamic sense because an asymmetric cloud will either break up or become more symmetric as time passes. Scale symmetry is typical of important processes believed to influence planet formation, such as turbulent flow within the solar nebula.

Nowadays we can look beyond the solar system. All hell breaks loose: the orbits of known exoplanets – planets round other stars – have all sorts of spacings, most of them very different from what we find in the solar system. On the other hand, the known exoplanets are an imperfect sample of those that actually exist; often only one planet is known for a given star, even though it probably has others. The detection methods are biased towards finding large planets circling close to their primaries.

Until we can map out the *entire* planetary systems of many stars, we won't really know what exoplanetary systems look like. But in 2013 Timothy Bovaird and Charles Lineweaver looked at 69 exoplanet systems known to have at least four planets, and 66 of them obey power laws. They also used the resulting power laws to tentatively predict 'missing' planets – doing a Ceres on an exosystem. Of the 97 planets predicted in this manner, only five have so far been observed. Even allowing for the difficulty of detecting small planets, this is a bit disappointing.

All of this is rather tentative, so attention has shifted to other principles that might explain how planetary systems are organised. These rely on subtle details of nonlinear dynamics and are not merely empirical. However, the patterns are less obviously numerical. In particular, Michael Dellnitz has shown mathematically that Jupiter's gravitational field seems to have arranged all of the other planets into an interconnected system linked by a natural set of 'tubes'. These tubes, which can be detected only through their mathematical features, provide natural low-energy routes between the different worlds. We'll discuss this idea along with related matters in Chapter 10, where it fits more naturally.

✦

Coincidence or not, the Titius–Bode law inspired some important discoveries.

The only planets visible to the naked eye are the classical five: Mercury, Venus, Mars, Jupiter, and Saturn. Plus Earth, if you want to be pedantic, but we only ever see a small part of it at one time. With the invention of the telescope, astronomers could observe stars that are too dim to see with the eye alone, along with other objects such as comets, nebulas, and satellites. Working at the limits of what was then technically feasible, the early astronomers often found it easier to spot a new object than to decide what it was.

Exactly this problem confronted William Herschel in 1781, when he pointed the telescope in the garden of his house in Bath towards the constellation Taurus and noticed a faint spot of light near the star zeta Tauri, which at first he thought was either 'a Nebulous Star or perhaps a Comet'. Four nights later he wrote in his journal that he had 'found it was a Comet, for it has changed its place'. About five weeks after that, when he reported his discovery to the Royal Society, he still described it as a comet. If you observe a star using lenses which magnify by different amounts, it remains pointlike even at the highest magnification, but this new object seemed to become larger as the magnification increased – 'as planets are', he remarked. But the same goes for comets, and Herschel was convinced he'd discovered a new comet.

As more information came in, some astronomers begged to differ, among them the Astronomer Royal Nevil Maskelyne, Anders Lexell, and Bode. By 1783 there was a consensus that the new object was a planet, and it required a name. King George III had given Herschel £200 a year on condition that he moved close enough to Windsor Castle for the royal family to look through his telescopes. Herschel, minded to repay him, wanted to call it Georgium Sidus, 'George's Star'. Bode suggested Uranus, the Latin form of Ouranos, the Greek sky god, and this name won the day, despite being the only planetary name based on a Greek god rather than a Roman one.

Laplace, quick off the mark, calculated the orbit of Uranus in 1783. The period is 84 years and the average distance from the Sun is about 19 AU or 3 billion kilometres. Although almost circular, Uranus's orbit is more eccentric than that of any other known planet, with a radius that ranges from 18 to 20 AU. Over the years, better telescopes made it possible to measure the planet's rotational period, which is 17 hours 14 minutes, and to reveal that it's retrograde – the planet spins in the opposite direction to every other. Its axis is tilted through more than a right angle, lying pretty much in the ecliptic plane of the solar system instead of being roughly perpendicular to it. As a result, Uranus experiences an extreme form of midnight sun: each pole endures 42 years of daylight followed by 42 of darkness, with one pole being dark while the other is light.

Clearly there's something strange about Uranus. On the other hand, it fits the Titius–Bode law perfectly.

Once the orbit was known and past sightings could be associated with the new world, it became apparent that it had been spotted before, but misidentified as a star or a comet. Indeed, it's just visible with good eyesight, and it was plausibly one of the 'stars' in Hipparchus's catalogue of 128 BC, and, later, in Ptolemy's *Almagest*. John Flamsteed observed it six times in 1690, thinking it was a star, then named 34 Tauri. Pierre Lemonnier observed it twelve times between 1750 and 1769. Although Uranus is a planet, it moves so slowly that it's easy not to notice any change in its position.

✦

So far, the main role of mathematics in understanding the solar system had been mainly descriptive, reducing lengthy series of observations to a simple elliptical orbit. The only prediction emerging from the mathematics was to forecast the planet's position in the sky at future dates. But, as time passed and enough observations accumulated, Uranus increasingly appeared to be in the wrong place. Alexis Bouvard, a student of Laplace, made numerous high-precision observations of Jupiter, Saturn, and Uranus, as well as discovering eight comets. His tables of the motion of Jupiter and Saturn proved to be very accurate, but Uranus drifted steadily away from its predicted location. Bouvard suggested that an even more distant planet might be perturbing Uranus's orbit.

'Perturb' here means 'have an effect on'. If we could express that effect mathematically in terms of the orbit of this presumptive new planet, we could work backwards to deduce that orbit. Then astronomers would know where to look, and if the prediction was based on fact, they could find the new planet. The big snag with this approach is that the motion of Uranus is influenced significantly by the Sun, Jupiter, and Saturn. The rest of the solar system's bodies can perhaps be neglected, but we still have five bodies to deal with. No exact formulas are known for three bodies; five are much harder.

Fortunately, the mathematicians of the day had already thought of a clever way to get round that issue. Mathematically, a perturbation of a system is a new effect that changes the solutions of its equations. For example, the movement of a pendulum under gravity *in a vacuum* has an elegant solution: the pendulum repeats the same oscillations over and again forever. If there's air resistance, however, the equation of motion changes to include this extra resistive force. This is a perturbation of the pendulum model, and it destroys the periodic oscillations. Instead, they die down and eventually the pendulum stops.

Perturbations lead to more complex equations, which are usually harder to solve. But sometimes the perturbation itself can be used to find out how solutions change. To do this we write down equations for the *difference* between the unperturbed solution and the perturbed one. If the perturbation is small, we can derive formulas that approximate this difference by neglecting terms in the equations that are much smaller than the perturbation. This trick simplifies the equations enough to solve them explicitly. The resulting solution is not exact, but it's often good enough for practical purposes.

If Uranus were the only planet, its orbit would be a perfect ellipse. However, this ideal orbit is perturbed by Jupiter, Saturn, and any other bodies of the solar system that we know about. Their combined gravitational fields change the orbit of Uranus, and this change can be described as a slow variation in the orbital elements of Uranus's ellipse. To a good approximation, Uranus always moves along *some* ellipse, but it's no longer always the *same* ellipse. The perturbations slowly change its shape and inclination.

In this manner we can calculate how Uranus would move when all important perturbing bodies are accounted for. The observations show that Uranus does not, in fact, follow this predicted orbit. Instead, it gradually deviates in ways that can be measured. So we add a hypothetical perturbation by an unknown Planet X, calculate the new perturbed orbit, set that equal to the observed one, and deduce the orbital elements of Planet X.

In 1843, in a computational *tour de force*, John Adams calculated the hypothetical new world's orbital elements. By 1845 Urbain Le Verrier was carrying out similar calculations independently. Adams sent his predictions to George Airy, the current British Astronomer Royal, asking him to look for the predicted planet. Airy was worried about some aspects of the calculation – wrongly, as it transpired – but Adams was unable to reassure him, so nothing was done. In 1846 Le Verrier published his own prediction, which again aroused little interest, until Airy noticed that both mathematicians had come up with very similar results. He instructed James Challis, Director of the Cambridge Observatory, to look for the new planet, but Challis failed to find anything.

Soon after, however, Johann Galle spotted a faint point of light about 1 degree away from Le Verrier's prediction and 12 degrees from Adams's. Later, Challis found that he'd observed the new planet twice, but didn't have an up-to-date star map and had generally been a bit sloppy, so he'd missed it. Galle's spot of light was another new planet, later named Neptune. Its discovery was a major triumph of celestial mechanics. Now mathematics was revealing the existence of unknown worlds, not just codifying the orbits of known ones.

✦

The solar system now boasted eight planets and a rapidly growing number of 'minor planets', or asteroids (see Chapter 5). But even prior to the discovery of Neptune, some astronomers, among them Bouvard and Peter Hansen, were convinced that a single new body couldn't explain the anomalies in the motion of Uranus. Instead, they believed that the discrepancies were evidence for *two* new planets. This idea was batted to and fro for another 90 years.

Percival Lowell founded an observatory in Flagstaff Arizona in 1894, and 12 years later he decided to sort out once and for all the anomalies in the orbit of Uranus, starting a project that he named Planet X. Here X is the mathematical unknown, not a Roman numeral (which would have been IX anyway). Lowell had somewhat ruined his scientific reputation by promoting the idea of 'canals' on Mars, and wanted to restore it: a new planet would be ideal. He used mathematical methods to predict where this hypothetical world should be, and then made a systematic search – with no result. He tried again in 1914–16, but again found nothing.

Meanwhile Edward Pickering, Director of Harvard College Observatory, had come up with his own prediction: Planet O, at a distance of 52 AU. By then the British astronomer Philip Cowell had declared the whole search a wild goose chase: the supposed anomalies in the motion of Uranus could be accounted for by other means.

In 1916 Lowell died. A legal dispute between his widow and the observatory put paid to any further search for Planet X until 1925, when Lowell's brother George paid for a new telescope. Clyde Tombaugh was given the job of photographing regions of the night sky twice, two weeks apart. An optical device compared the two images, and anything that had changed position would blink, drawing attention to the movement. He took a third image to resolve any uncertainties. Early in 1930, he was examining an area in Gemini and something blinked. It was within 6 degrees of a location suggested by Lowell, whose prediction appeared to be upheld. Once the object was identified as a new planet, a search of the archives showed that it had been photographed in 1915 but not then recognised as a planet.

The new world was dubbed Pluto, its first two letters being Lowell's initials.

Pluto turned out to be much smaller than expected, with a mass only one tenth that of the Earth. That implied that it could not, in fact, explain the anomalies that had led Lowell and others to predict its existence. When the low mass was confirmed in 1978, a few astronomers resumed the search for Planet X, believing that Pluto was a red herring and a more massive unknown planet must be out there somewhere. When Myles Standish used data from the 1989 Voyager fly-by of Neptune to refine the figure for Neptune's mass, the anomalies in the orbit of Uranus vanished. Lowell's prediction was just a lucky coincidence.

Pluto is weird. Its orbit is inclined at 17 degrees to the ecliptic, and is so eccentric that for a time Pluto comes closer to the Sun than Neptune. However, there's no chance of them colliding, for two reasons. One is the angle between their orbital planes: their orbits cross only on the line where those planes meet. Even then, both worlds must pass through the same point on this line at the same time. This is where the second reason comes in. Pluto is locked into a 2:3 resonance with Neptune. The two bodies therefore repeat essentially the same movements every two orbits of Pluto and three of Neptune, that is, every 495 years. Since they haven't collided in the past, they won't do so in the future – at least, not until large-scale reorganisation of other bodies of the solar system disturbs their cosy relationship.

✦

Astronomers continued searching the outer solar system for new bodies. They discovered that Pluto has a comparatively large moon, Charon, but nothing else was spotted beyond the orbit of Neptune until 1992, when a small body christened (15760) 1992 QB_1 showed up. It was so obscure that this is still its name (a proposal to call it 'Smiley' was rejected because that name had already been used for an asteroid), but it proved to be the first of a gaggle of trans-Neptunian objects (TNOs), of which more than 1500 are known. Among them are a few larger bodies, though still smaller than Pluto: the largest is Eris, followed by Makemake, Haumea, and 2007 OR$_{10}$.

All of these objects are too lightweight and too distant to be predicted from their gravitational effects on other bodies, and were discovered by searching through images. But there are some noteworthy mathematical features, related to the effects of other bodies on *them*. Between 30 and 55 AU lies the Kuiper belt, most of whose members are in roughly circular orbits close the ecliptic. Some of these TNOs are in resonant orbits with Neptune. Those in 2:3 resonance are called plutinos, because they include Pluto. Those in 1:2 resonance – period twice that of Neptune – are called twotinos. The rest are classical Kuiper belt objects, or cubewanos;[4] they also have roughly circular orbits, but experience no significant perturbations from Neptune. Further out is the scattered disc. Here, asteroid-like bodies move in eccentric orbits, often inclined at a large angle to the ecliptic. Among them are Eris and Sedna.

As more and more TNOs were found, some astronomers began to feel that it made little sense to call Pluto a planet, but not Eris, which they thought was slightly bigger. Ironically, images from *New Horizons* showed that Eris is slightly smaller than Pluto.[5] But once other TNOs were classified as planets, some would be smaller than the asteroid (or minor planet) Ceres. After much heated debate, the International Astronomical Union demoted Pluto to the status of dwarf planet, where it was joined by Ceres, Haumea, Makemake, and Eris. New definitions of the terms 'planet' and 'dwarf planet' were carefully tailored to shoehorn the bodies concerned into those two classifications. However, it's not yet clear whether Haumea, Makemake, and Eris actually fit the definition. It's also suspected that a few hundred further dwarf planets exist in the Kuiper belt, and up to ten thousand in the scattered disc.

✦

When a new scientific trick works, it's only sensible to try it on similar problems. Used to predict the existence and location of Neptune, the perturbation trick worked brilliantly. Tried on Pluto, it seemed to work brilliantly as well, until astronomers realised that Pluto is too small to create the anomalies used to predict it.

The trick failed dismally for a planet named Vulcan. This is not the fictional planet of *Star Trek*, homeworld of Mr Spock, which according to the science fiction writer James Blish orbits the star 40 Eridani A. Instead, it's the fictional planet that orbits an obscure and rather ordinary star known to science fiction writers as Sol. Or, more familiarly, the Sun. Vulcan teaches us several lessons about science: not just the obvious one that mistakes can be made, but the more general point that awareness of past mistakes can stop us repeating them. Its prediction is linked to the introduction of relativity as an improvement on Newtonian physics. But more of that story later, as they say.

Neptune was discovered because of anomalies in the orbit of Uranus. Vulcan was proposed to explain anomalies in the orbit of Mercury – and the proposer was none other than Le Verrier, in work that predates Neptune. In 1840 the Director of the Paris Observatory, François Arago, wanted to apply Newtonian gravitation to the orbit of Mercury, and asked Le Verrier to perform the necessary calculations. When Mercury passed across the face of the Sun, an event called a transit, the theory could be tested by observing the times when the transit began and ended. There was a transit in 1843, and Le Verrier completed his calculations shortly beforehand, making it possible to predict the timing. To his dismay, the observations failed to agree with the theory. So Le Verrier went back to the drawing board, preparing a more accurate model based on numerous observations and 14 transits. And by 1859 he had noticed, and published, a small but baffling aspect of Mercury's motion that explained his original error.

The point at which Mercury's orbital ellipse gets nearest to the Sun, known as the perihelion (*peri* = close, *helios* = Sun), is a well-defined feature. As time passes, Mercury's perihelion slowly rotates relative to the background of distant ('fixed') stars. In effect, the entire orbit slowly pivots with the Sun at its focus; the technical term is precession. A mathematical result known as Newton's theorem of revolving orbits[6] predicts this effect as a consequence of perturbations by other planets. However, when Le Verrier plugged the observations into this theorem, the resulting numbers were very slightly wrong. Newtonian theory predicted that the perihelion of Mercury should precess through 532″ (seconds of arc) every hundred years; the observed figure was 575″. Something was causing an extra 43″ precession per century. Le Verrier suggested that some undiscovered planet, orbiting closer to the

Sun than Mercury, was responsible, and he named it Vulcan, after the Roman god of fire.

The Sun's glare would overwhelm any light reflected from such a closely orbiting planet, so the only practical way to observe Vulcan would be during a transit. Then it should be visible as a tiny dark dot. The amateur astronomer Edmond Lescarbault quickly announced that he'd found such a dot, which wasn't a sunspot because it moved at the wrong speed. Le Verrier announced the discovery of Vulcan in 1860, and on the strength of this he was awarded the Legion of Honour.

Unfortunately for Le Verrier and Lescarbault, a better-equipped astronomer, Emmanuel Liais, had also been observing the Sun at the behest of the Brazilian government, and had seen nothing of the kind. His reputation was at stake, and he denied that any such transit had occurred. The arguments became heated and confused. When Le Verrier died in 1877 he still believed he'd discovered another planet. Without Le Verrier's backing, the Vulcan theory lost momentum, and soon the consensus was straightforward: Lescarbault had been wrong. Le Verrier's prediction remained unverified, and there was widespread skepticism. Interest vanished almost totally in 1915, when Einstein used his new theory of general relativity to derive a precession of $42 \cdot 98''$ without any assumption of a new planet. Relativity was vindicated and Vulcan was thrown on the scrapheap.

We still don't know for certain that there are no bodies between Mercury and the Sun, though if one exists, it has to be very small. Henry Courten reanalysed images of the solar eclipse of 1970, stating that he'd detected at least seven such bodies. Their orbits couldn't be determined, and the claims haven't been confirmed. But the search for vulcanoids, as they're named, continues.[7]

Celestial Police

> The dinosaurs didn't have a space program, so they're not here
> to talk about this problem. We are, and we have the power to do
> something about it. I don't want to be the embarrassment of the
> galaxy, to have had the power to deflect an asteroid, and then not,
> and end up going extinct.
>
> Neil deGrasse Tyson, *Space Chronicles*

PURSUED BY A FLEET of interstellar warships firing sizzling bolts of
pure energy, a small band of courageous freedom fighters seeks refuge
in an asteroid belt, weaving violently through a blizzard of tumbling
rocks the size of Manhattan that constantly smash into each other.
The warships follow, evaporating the smaller rocks with laser beams
while accepting numerous hits from smaller fragments. In a cunning
manoeuvre, the fleeing vessel loops back on itself and dives into a deep
tunnel at the centre of a crater. But its worries have only begun…

It's a breathtaking cinematic image.

It's also total nonsense. Not the fleet of warships, the energy bolts,
or the galactic rebels. Not even the monstrous worm lurking at the end
of the tunnel. Those *might* just happen one day. It's that blizzard of
tumbling rock. No way.

I reckon it's all down to that badly chosen metaphor. Belt.

✦

Once upon a time the solar system, as then understood, lacked a belt.
Instead, there was a gap. According to the Titius–Bode law, there ought

to have been a planet between Mars and Jupiter, but there wasn't. If there had been, the ancients would have seen it and associated yet another god with it.

When Uranus was discovered, it fitted so neatly into the mathematical pattern of the Titius–Bode law that astronomers were encouraged to fill the gap between Mars and Jupiter. As we saw in the previous chapter, they succeeded. Baron Franz Xaver von Zach initiated the *Vereinigte Astronomische Gesellschaft* (United Astronomical Society) in 1800, with 25 members – among them Maskelyne, Charles Messier, William Herschel, and Heinrich Olbers. Because of its dedication to tidying up the unruly solar system, the group became known as the *Himmelspolizei* (celestial police). Each observer was assigned a 15-degree slice of the ecliptic and tasked with searching that region for the missing planet.

As is all too common in such matters, this systematic and organised approach was trumped by a lucky outsider: Giuseppe Piazzi, Astronomy Professor at the University of Palermo in Sicily. He wasn't searching for a planet; he was looking for a star, 'the 87th of the Catalogue of Mr. La Caille'. Early in 1801, near the star he was seeking, he saw another point of light, which matched nothing in the existing star catalogues. Continuing to observe this interloper, he found that it moved. His discovery was exactly where the Titius–Bode law required it to be. He named it Ceres after the Roman harvest goddess, who was also the patron goddess of Sicily. At first he thought he'd spotted a new comet, but it lacked the characteristic coma. 'It has occurred to me several times that it might be something better than a comet,' he wrote. Namely, a planet.

Ceres is rather small by planetary standards, and astronomers nearly lost it again. They had very little data on its orbit, and before they could obtain more measurements the motion of the Earth carried the line of sight to the new body too close to the Sun, so its faint light was swamped by the glare. It was expected to reappear a few months later, but the observations were so sparse that the likely position would be very uncertain. Not wishing to start the search all over again, astronomers asked the scientific community to provide a more reliable prediction. Carl Friedrich Gauss, then a relative unknown in the public eye, rose to the challenge. He invented a new way to deduce an orbit from three or more observations, now known as Gauss's method. When

Ceres duly reappeared within half a degree of the predicted position, Gauss's reputation as a great mathematician was sealed. In 1807 he was appointed Professor of Astronomy and Director of the Observatory at Göttingen University, where he remained for the rest of his life.

To predict where Ceres would reappear, Gauss invented several important numerical approximation techniques. Among them was a version of what we now call the fast Fourier transform, rediscovered in 1965 by James Cooley and John Tukey. Gauss's ideas on the topic were found among his unpublished papers and appeared posthumously in his collected works. He viewed this method as a form of trigonometric interpolation, inserting new data points between existing ones in a smooth manner. Today it's a vital algorithm in signal processing, used in medical scanners and digital cameras. Such is the power of mathematics, and what the physicist Eugene Wigner called its 'unreasonable effectiveness'.[1]

Building on this success, Gauss developed a comprehensive theory of the motion of small asteroids perturbed by large planets, which appeared in 1809 as *Theoria Motus Corporum Coelestium in Sectionibus Conicis Solem Ambientum* (Theory of Motion of Celestial Bodies Moving in Conic Sections around the Sun). In this work, Gauss refined and improved a statistical method introduced by Legendre in 1805, now called the method of least squares. He also stated that he'd had the idea first, in 1795, but (typically for Gauss) he hadn't published it. This method is used to deduce more accurate values from a series of measurements, each subject to random errors. In its simplest form it selects the value that minimises the total error. More elaborate variations are used to fit the best straight line to data about how one variable relates to another, or to deal with similar issues for many variables. Statisticians use such methods on a daily basis.

✦

When the orbital elements of Ceres were safely in the bag, so that it could be found whenever required, it turned out not to be alone. Similar bodies, of similar size or smaller, had similar orbits. The better your telescope, the more of them you could see, and the smaller they became.

Later in 1801 one of the Celestial Police, Olbers, spotted one such body, naming it Pallas. He quickly came up with an ingenious explanation for the absence of one large planet but the presence of two (or more). There had once been a large planet in this orbit, but it had broken up in a collision with a comet or a volcanic explosion. For a time this idea seemed plausible, because more and more 'fragments' were found: Juno (1804), Vesta (1807), Astraea (1845), Hebe, Iris, and Flora (1847), Metis (1848), Hygeia (1849), Parthenope, Victoria, and Egeria (1850), and so on

Vesta can sometimes be seen with the naked eye, under favourable observing conditions. The ancients could have discovered it.

Traditionally, each planet had its own symbol, so initially each of the newly discovered bodies was also given an arcane symbol. As new bodies flooded in, this system proved too cumbersome, and it was replaced by more prosaic ones, evolving into today's – basically a number indicating order of discovery, a name or temporary designation, and a date of discovery, such as 10 Hygeia 1849.

In a sufficiently powerful telescope, a planet looks like a disc. These objects were so small that they appeared as points, just like stars. In 1802 Herschel suggested a working name for them:

> They resemble small stars so much as hardly to be distinguished
> from them. From this, their asteroidal appearance, if I take my name,
> and call them Asteroids; reserving for myself however the liberty
> of changing that name, if another, more expressive of their nature,
> should occur.

For a while, lots of astronomers continued to call them planets, or minor planets, but eventually 'asteroid' prevailed.

Olbers's theory has not survived the test of time. The chemical make-up of the asteroids is not consistent with them all having been fragments of a single larger body, and their combined mass is far too small. They're more likely to be cosmic debris left over from a would-be planet that failed to form because Jupiter caused too much disturbance. Collisions between planetesimals in this region were more common than elsewhere, and broke them up faster than they could aggregate. This was caused by Jupiter's migration towards the Sun, mentioned in Chapter 1.

The problem wasn't Jupiter as such, but resonant orbits. These occur, as previously remarked, when the period of a body pursuing one orbit is a simple fraction of that of another body – here Jupiter. Both bodies then follow a cycle in which they end up in exactly the same relative positions as they occupied at the start. And this *keeps happening*, causing a large disturbance. If the ratio of the periods is not a simple fraction, such effects get smeared out. Exactly what happens depends on the fraction, but there are two main possibilities. Either the distribution of asteroids near that orbit is concentrated, so there are more of them than is typical elsewhere, or they're cleared out from that orbit altogether.

If Jupiter stayed in the same orbit, this process would eventually settle down as asteroids concentrated near stable resonances and avoided unstable ones. But if Jupiter started to move, as astronomers now believe it did, the resonance zones would sweep across the asteroid belt, causing mayhem. Before anything could settle down in a nice stable resonance, the orbit concerned would stop being resonant and everything would be disturbed again. The movement of Jupiter therefore churned up the asteroids, keeping their dynamics erratic and increasing the chances of a collision. The inner planets are evidence that planetesimals aggregated inside the orbits of the giant planets, implying that lots of planetesimals once existed. Several giants would be likely to perturb each other, much as Jupiter and Saturn did, which would change their orbits; changing orbits implies sweeping resonance zones, which break up any planetesimals that are just inside the orbit of the innermost giant. In short, inner planets plus two or more giants imply asteroids.

✦

Belt.

As far as I can determine, no one knows precisely who first used the term 'asteroid belt', but it was definitely in use by 1850 when Elise Otté's translation of Alexander von Humboldt's *Cosmos*, discussing meteorite showers, remarked that some of these 'probably form part of a belt of asteroids intersecting the Earth's orbit'. Robert Mann's 1852 *A Guide to the Knowledge of the Heavens* states: 'The orbits of the asteroids are placed in a wide belt of space.' And so they are. The picture shows the distribution of the principal asteroids, along with the orbits of the planets out to Jupiter. A huge fuzzy ring composed of thousands of asteroids dominates the picture. I'll come back to the Hildas, Trojans, and Greeks later.

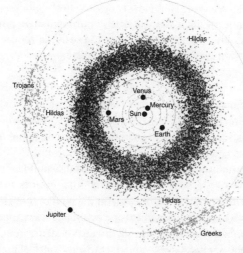

The asteroid belt, along with three major clumps of asteroids: Hildas, Trojans, and Greeks. Drawn in a rotating frame so that Jupiter remains fixed.

This image, reinforced by the term 'belt', is probably why the *Star Wars* movies – and, what's worse, popular science shows on TV, whose producers really should know better – routinely depict asteroids as a swarm of closely packed rocks, continually smashing into each other. It makes for exciting cinema, but it's nonsense. Yes, there are lots of rocks out there … but there's also a truly enormous amount of *space* out there. A back-of-the-envelope calculation shows that the typical distance between asteroids 100 metres or more across is about 60,000 kilometres. That's roughly five times the diameter of the Earth.[2] So, despite Hollywood movies, if you were in the asteroid belt you wouldn't see hundreds of rocks floating around. You probably wouldn't see anything.

The real problem is that fuzzy image. In a diagram, with dots for the various bodies, the asteroids form a dense stippled ring. So we expect the real thing to be equally dense. But each dot in the image represents a region of space roughly *three million kilometres* across. The same goes for similar features of the solar system. The Kuiper belt is not a belt, and the Oort cloud is not a cloud. Both are almost entirely empty space. But it's such a *lot* of space that the tiny proportion that isn't space consists of truly vast numbers of celestial bodies – mainly rock and ice. We'll come to those two regions later.

✦

Finding patterns in data is something of a black art, but mathematical techniques can help. One basic principle is that different ways to present or plot data can bring out different features.

The illustration suggests that asteroids are fairly uniformly scattered within the main asteroid belt. The ring of dots looks much the same density everywhere, with no holes. But once again the picture is misleading. Its scale is too compressed to show the true details, but more importantly it shows the *positions* of asteroids. To see interesting structure – other than the two clusters labelled Greeks and Trojans, to which I'll return – we have to look at distances. In fact, what really counts is rotational periods, but those are related to distance by Kepler's third law.

In 1866 an amateur astronomer named Daniel Kirkwood noticed gaps in the asteroid belt. More precisely, asteroids rarely orbit at certain distances from the Sun, as measured by the major radius of the orbital ellipse. The picture shows a modern and more extensive plot of the number of asteroids at a given distance, in the core of the belt which ranges from 2 to 3·5 AU. Three sharp dips, where the number of asteroids goes to zero, are evident. Another one occurs near 3·3 AU, but it's not so obvious because there are fewer asteroids that far out. These dips are called Kirkwood gaps.

The Kirkwood gaps are not apparent in this picture for two reasons. The pixels representing asteroids are large compared to the size of the asteroids on the scale of the image, and the 'gaps' occur in distances, not location. Each asteroid follows an elliptical orbit, and its distance from the Sun varies around the orbit. So asteroids *cross* the gaps; they just don't stay there for long. The major axes of these ellipses point in many different directions. These effects make the gaps so fuzzy that they can't be seen in a picture. Plot the distances, however, and they immediately stand out.

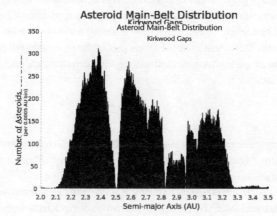

Kirkwood gaps in the asteroid belt and associated resonances with Jupiter.

Kirkwood suggested, correctly, that the gaps are caused by Jupiter's massive gravitational field. This affects every asteroid in the belt, but there's a significant difference between resonant orbits and non-resonant ones. The very deep dip at the left of the picture corresponds to an orbital distance at which the asteroid is in 3:1 resonance with Jupiter. That is, it makes three revolutions of the Sun for every one made by Jupiter. This repeated alignment makes the long-term effect of Jupiter's gravity stronger. In this case the resonances clear out regions of the belt. The orbits of asteroids in resonance with Jupiter become more elongated and chaotic, to such an extent that they cross the orbits of inner planets, mainly Mars. Occasional closer encounters with Mars change the orbits even more, flinging them in random directions. As this effect causes the zone near the resonant orbit to lose ever more asteroids, gaps are created there.

The main gaps, with the corresponding resonances in parentheses, are at distances of 2·06 AU (4:1), 2·5 AU (3:1), 2·82 AU (5:2), 2·95 AU (7:3), and 3·27 AU (2:1). There are weaker or narrower ones at 1·9 AU (9:2), 2·25 AU (7:2), 2·33 AU (10:3), 2·71 AU (8:3), 3·03 AU (9:4), 3·08

AU (11:5), 3·47 AU (11:6), and 3·7 AU (5:3). So resonances control the distribution of the major radii of the asteroids.

As well as gaps, there are clumps. Again, this term usually refers to concentrations near a given distance, not actual local clusters of asteroids. However, we'll come to two genuine clusters next, the Greeks and Trojans. Resonances sometimes cause clusters rather than gaps, depending on the numbers that occur in the resonance and various other factors.[3]

✦

Even though the general three-body problem – how three point masses move under Newtonian gravitation – is extremely difficult to solve mathematically, some useful results can be obtained by focusing on special solutions. Paramount among them is the '2½-body problem', a mathematical joke with a serious point to it. In this set-up, two of the bodies have non-zero mass but the third body is so tiny that its mass is effectively zero. An example is a speck of dust moving under the influence of the Earth and the Moon. The idea behind the model is that the dust speck responds to gravitational forces exerted by the Earth and the Moon, but it's so light that it effectively exerts no force on either body. Newton's law of gravity tells us that the dust grain would actually exert a very small force, but this is so small that it can be ignored in the model. In practice, the same is true for a heavier body too, such as a small moon or asteroid, provided the timescale is short enough to exclude significant chaotic effects.

There's one further simplification: the two large bodies move in circular orbits. This lets us transform the entire problem into a rotating frame of reference, relative to which the large bodies are stationary and lie in a fixed plane. Imagine a huge turntable. Attach the Earth and Moon to the turntable so that they lie on a straight line that passes through the central pivot, on opposite sides of it. The Earth's mass is about 80 times that of the Moon; if we place the Moon 80 times as far from the pivot as the Earth is, the centre of mass of the two bodies coincides with the pivot. If the turntable now spins at just the right speed, carrying the Earth and Moon with it, they follow circular orbits consistent with Newtonian gravity. Relative to a system of cooordinates attached to the turntable, both bodies are stationary, but they experience the rotation as 'centrifugal force'. This isn't a genuine physical force: it arises because the bodies are glued to the turntable and can't move in straight lines. But it has the same effect on their dynamics as a force in that system of coordinates. For this reason it's often referred to as a 'fictitious force', even though its effect is real.

In 1765 Euler proved that in such a model you can glue a dust particle to a point on the same straight line as the other two bodies, so that all three bodies move in circular orbits consistent with Newtonian gravity. At such a point, the gravitational forces exerted by the Earth and the Moon are exactly cancelled by the centrifugal force experienced by the dust speck. In fact, Euler found three such points. One of them, now called L1, lies between the Earth and the Moon. L2 lies on the far side of the Moon as seen from the Earth; L3 lies on the far side of the Earth as seen from the Moon.

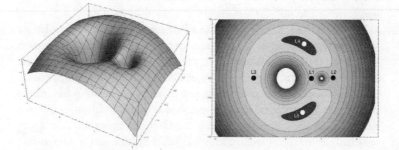

Gravitational landscape for the 2½-body problem in a
rotating frame. Left: the surface. Right: its contours.[4]

These symbols use the letter L instead of E because in 1772 Lagrange found two more possible locations for the dust speck. These don't lie on the Earth–Moon line, but at the corners of two equilateral triangles, whose other two corners are the Earth and the Moon. At these points, the dust speck remains stationary relative to the Earth and the Moon. The Lagrange point L4 is 60 degrees ahead of the Moon and L5 is 60 degrees behind it. Lagrange proved that, for any two bodies, there are precisely five such points.

Technically the orbits corresponding to L4 and L5 generally have different radii from those of the other two bodies. However, if one of those is much more massive – for instance, when it's the Sun and the other body is a planet – then the common centre of mass and the more massive body almost coincide. The orbits corresponding to L4 and L5 are then almost the same as that of the less massive body.

The geometry of the Lagrange points can be worked out from the energy of the dust speck. This consists of its kinetic energy as it rotates along with the turntable, plus its gravitational potential energies corresponding to attraction by the Earth and the Moon. The picture shows the total energy of the dust speck in two ways: as a curved surface, whose height represents the total energy, and as a system of contours, curves on which the energy is constant. You can think of the surface as a kind of gravitational landscape. The speck moves across the landscape, but unless some force disturbs it, conservation of energy implies that it must stay on a single contour line. It can move sideways, but not up or down.

If the contour 'line' is a single point, the speck will be in equilibrium – it stays where you put it, relative to the turntable. There are five such points, marked on the contour picture as L1–L5. At L1, L2, and L3 the surface is shaped like a saddle: in some directions the landscape curves up, in others down. In contrast, both L4 and L5 are peaks in the energy landscape. The important difference is that peaks (and valleys, which don't occur here) are surrounded by small closed contours, which stay very close to the peak itself. Saddles are different: contours near saddle points head off into the distance, and while they may eventually close up, they make a big excursion first.

If the speck is slightly displaced, it moves a small distance away and then follows whichever contour line it lands on. For a saddle, all such contours take it far away from its original location. For example, if the speck starts at L2 and moves a little to the right, it lands on a huge closed contour that takes it all the way round the Earth, outside L3 on the far side. So saddle equilibria are *unstable*: the initial disturbance grows much larger. Peaks and valleys are *stable*: nearby contours are closed and *stay* nearby. A small initial disturbance remains small. The speck is no longer in equilibrium, however: its actual motion is a combination of a small oscillation around the closed contour and the overall rotation of the turntable. Such motion is called a tadpole orbit.

The key point is: the dust speck stays close to the peak.

(I've cheated a little here, because the picture shows positions but not velocities. Changes in velocity make the actual motion more complicated, but the stability results remain valid. See Chapter 9.)

Lagrange points are special features in the gravitational landscape, which can be exploited in planning space missions. In the 1980s there was a spate of interest in space colonies: giant artificial habitats for humans to live in, growing their own food, powered by sunlight. People could live on the interior of a hollow cylinder if it rotated about its axis, creating artificial gravity by centrifugal force. A Lagrange point is an attractive location, because it's an equilibrium. Even at one of those unstable saddles at L1, L2, or L3, an occasional burst from a rocket motor would stop the habitat drifting away. The peaks, L4 and L5, are even better – no such correction is needed.

✦

Nature knows about Lagrange points too, in the sense that there are real configurations similar enough to those considered by Euler and Lagrange for their results to work. Often these real examples violate some of the technical conditions in the model; for example, the dust particle need not lie in the same plane as the other two bodies. The main features of the Lagrange points are relatively robust, and continue to hold for anything reasonably similar to the idealised model.

The most spectacular instance is Jupiter, with its own space colonies: asteroids known as Trojans and Greeks. The picture is drawn at a specific time in a rotating frame that follows Jupiter round its orbit. Max Wolf found the first, 588 Achilles, in 1906. Some 3898 Greeks and 2049 Trojans were known in 2014. It's thought that about a million Greeks and Trojans larger than a kilometre across exist. The names are traditional: Johann Palisa, who calculated many of their orbital elements, suggested naming these bodies after participants in the Trojan war. Nearly all of the Greeks are near L4, and most Trojans are near L5. However, by an accident of history, the Greek Patroclus is placed among the Trojans and the Trojan Hektor is hemmed in by Greeks. Although these bodies form relatively small clusters in the picture, astronomers think there are about as many of them as there are regular asteroids.

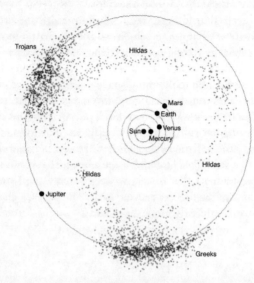

The Greek and Trojan asteroids form clumps. The Hilda family forms
a fuzzy equilateral triangle with two vertices at L4 and L5.

Greeks follow much the same orbit as Jupiter, but 60 degrees ahead
of it; Trojans are 60 degrees behind. As explained in the previous section,
the orbits are not identical to that of Jupiter; just close. Moreover, the
approximation of circular orbits in the same plane is unrealistic; many
of these asteroids have orbits inclined at up to 40 degrees to the ecliptic.
The clusters remain bunched because the L4 and L5 points are stable in
the 2½-body model, and Jupiter's large mass keeps them fairly stable
in the real many-body dynamics, because perturbations from elsewhere
– mainly Saturn – are relatively small. However, in the longer run a few
asteroids may be gained or lost by either cluster.

For similar reasons, other planets can also be expected to have their
own Trojans (for generic terminology Jupiter's Greeks are considered
honorary Trojans). Venus has a temporary one, 2013 ND$_{15}$. Earth has
a more permanent Trojan 2010 TK$_7$ at its L4 point. Mars boasts five,
Uranus has one, and Neptune has at least 12 – probably more than
Jupiter, perhaps ten times as many.

What of Saturn? No Trojan asteroids are known, but it has two Trojan satellites – the only known ones. Its moon Tethys has two Trojan moons of its own, Telesto and Calypso. Another of Saturn's moons, Dione, also has two Trojan moons, Helene and Polydeuces.

Jupiter's Trojans are closely associated with another fascinating group of asteroids, the Hilda family. These are in 3:2 resonance with Jupiter, and in a rotating frame they occupy a region shaped roughly like an equilateral triangle with vertices at L4, L5, and a point in Jupiter's orbit that is diametrically opposite the planet. The Hildas 'circulate' slowly relative to the Trojans and Jupiter.[5] Unlike most asteroids, they have eccentric orbits. Fred Franklin has suggested that the current orbits provide extra evidence that Jupiter initially formed about 10% further from the Sun and migrated inwards.[6] Asteroids at that distance with circular orbits would either have been cleared out as Jupiter migrated inwards, or would have changed to more eccentric orbits.[7]

The Planet that Swallowed its Children

The star of Saturn is not a single star, but is a composite of three, which almost touch each other, never change or move relative to each other, and are arranged in a row along the zodiac, the middle one being three times larger than the lateral ones, and they are situated in this form: oOo.

Galileo Galilei, Letter to Cosimo de' Medici, 30 July 1610

WHEN GALILEO FIRST POINTED his telescope at Saturn, and drew what he saw, it looked like this:

Galileo's 1610 drawing of Saturn.

You can see why he described it as oOo in his excited letter to his sponsor Cosimo de' Medici. He sent news of his discovery to Kepler, but as was common at the time he wrote it in the form of an anagram: *smaismrmilmepoetaleumibunenugttauiras*. If anyone later made the same discovery, Galileo could then claim priority by deciphering this as *Altissimum planetam tergeminum observavi*: 'I have observed the most distant planet to have a triple form.'

Unfortunately Kepler deciphered it as *Salve umbistineum geminatum Martia proles*: 'Be greeted, double knob, children of Mars.' That is, Mars has two moons. Kepler had already predicted this on the grounds that Jupiter has four and Earth one, so in between there ought to be two, because 1, 2, 4 is a geometric series. And presumably Saturn would have eight moons. Half a moon for Venus? Sometimes Kepler's ability to spot patterns was rather forced. But I should reserve my scorn, because, miraculously, Mars *does* have two moons, Phobos and Deimos.

When Galileo looked again in 1616, he realised that his rudimentary telescope had deceived him with a blurred image, which he had interpreted as three discs. But it was still rather baffling. Galileo wrote that Saturn seemed to have ears.

Galileo's 1616 drawing of Saturn.

A few years later, when he looked again, the ears or moons or whatever they were had vanished. Half in jest, Galileo wondered whether Saturn had eaten its children. This was an oblique reference to a rather gory Greek myth in which the Titan Kronos, terrified that one of his children would overthrow him, ate each of them when it was born. The Roman equivalent of Kronos is Saturn.

When the ears returned, Galileo was even more puzzled.

Of course, we now know the reality behind Galileo's observations. Saturn is surrounded by a gigantic system of circular rings. They're tilted relative to the ecliptic, so that as Saturn revolves round the Sun we sometimes see the rings 'full face', and they appear to be larger than the planet, like the 'ears' drawing. At other times we see them edge-on, and they vanish altogether unless we use a telescope a lot better than Galileo's.

This fact alone tells us that the rings are very thin compared to the planet, but we now know that they're very thin indeed, a mere 20 metres. Their diameter, in contrast, is 360,000 kilometres. If Saturn's rings were as thick as a pizza they'd be about the size of Switzerland. Galileo knew none of that. But he did know that Saturn was strange, mysterious, and quite unlike any other planet.

✦

Christiaan Huygens had a better telescope, and in 1655 he wrote that Saturn 'is surrounded by a thin, flat, ring, nowhere touching, inclined to the ecliptic'. Hooke even spotted shadows, both of the globe upon the ring, and the ring upon the globe, which clarify the three-dimensional geometry by showing us what's in front of what.

Left: Hooke's 1666 drawing of Saturn, showing shadows. Right: Modern image showing the Cassini division, a prominent dark gap in the rings.

Are Saturn's rings solid, like the brim of a hat, or are they composed of myriad tiny lumps of rock or ice? If solid, what material are they made of? If not, why do the rings appear to be rigid, their shape unchanging?

Answers came bit by bit, in a mixture of observations and mathematical analysis.

Early observers saw a single broad ring. In 1675, however, Giovanni Cassini obtained better observations that revealed a number of circular gaps, dividing the overall ring into a series of concentric subrings. The most prominent gap is known as the Cassini division. The innermost ring is the B ring, the outermost is the A ring. Cassini also knew of a fainter C ring inside the B ring. These discoveries deepened the mystery, but paved the way to its eventual solution.

Laplace, in 1787, pointed out a dynamic problem with a wide solid ring. Kepler's third law tells us that the more distant a body is from the centre of a planet, the more slowly it revolves. But the inner and outer edges of a solid ring revolve at the same angular speed. So either the outer edge is moving too quickly, or the inner one is moving too slowly, or both. This discrepancy sets up stresses in the material of the ring, and unless it's made of amazingly strong material, it will break apart. Laplace's solution to this difficulty was elegant: he suggested that the rings are composed of large numbers of very narrow ringlets, fitting one inside the next. Each ringlet is solid, but their speeds of revolution decrease as their radii get bigger. This neatly sidestepped the stress problem, because the inner and outer edges of a narrow ring revolve at almost the same speed.

Elegant, but misguided. In 1859 the mathematical physicist James Clerk Maxwell proved that a rotating solid ringlet is unstable. Laplace had solved the problem of stress caused by the edges rotating at different speeds, but these were 'shear' stresses, like the forces between cards in a pack if you slide the cards sideways while keeping them in a pile. But other stresses can come into play – for example, bending the cards. Maxwell proved that for a solid ringlet, any slight disturbances caused by external sources grows, causing the ringlet to ripple and bend, so it breaks up, like a dry stick of spaghetti which snaps as soon as you try to bend it.

Maxwell deduced that Saturn's rings must be composed of innumerable tiny bodies, all moving independently in circles, at whichever speed is mathematically consistent with the gravitational pull exerted on them. (Recently some problems with this type of simplified model have become apparent: see Chapter 18. The implication for ring models is unclear. I'll postpone further discussion to that chapter and report the conventional results here.)

Because everything moves in circles, the entire set-up is rotationally symmetric, so the speed of a particle depends only on its distance from the centre. Assuming the mass of the ring material to be negligible compared to that of Saturn, which we now know is the case, Kepler's third law leads to a simple formula. The speed of a ring particle in kilometres per second is 29·4 divided by the square root of its orbital radius, measured as a multiple of the radius of Saturn.

Alternatively, the rings might be liquid. But in 1874 Sofia Kovalevskaya, one of the great women mathematicians, showed that a liquid ring is also unstable.

By 1895 the verdict of observational astronomers was in. Saturn's rings are composed of a vast number of small bodies. Further observations led to several new, even fainter, subrings, imaginatively named D, E, F, and G. These were named in order of discovery, and their spatial order, reading outwards from the planet, is DCBAFGE. Not quite as jumbled as Galileo's anagram, but getting there.

✦

No military plan survives contact with the enemy. No astronomical theory survives contact with better observations.

In 1977 NASA sent two space probes, *Voyager 1* and *2*, on a planetary grand tour. The planets of the solar system had fortuitously aligned themselves so that it was possible to visit the outer planets in turn. *Voyager 1* visited Jupiter and Saturn; *Voyager 2* also passed by Uranus and Neptune. The *Voyager*s continued their travels, heading off into interstellar space, defined as the region beyond the heliopause, where the solar wind runs out of steam. So 'interstellar' means that the Sun no longer has any significant influence beyond a very weak gravitational pull. *Voyager 1* reached this transition zone in 2012, and *Voyager 2* is expected to do the same in 2016. Both continue to send back data. They have to be the most successful space missions ever.

Late in 1980, humanity's ideas about Saturn changed forever when *Voyager 1* started sending back pictures of the rings, six weeks before its closest approach to the planet. Fine details never seen before showed that there are hundreds, if not thousands, of distinct rings, very closely spaced like the grooves on an old gramophone record. That in itself wasn't a huge surprise, but other features were unexpected and at first rather baffling. Many theoreticians had expected the main features of the ring system to coincide with resonances with the planet's innermost (known) satellites, but on the whole they don't. The Cassini division turned out not to be empty; there are at least four thin rings inside it.

Rich Terrile, one of the scientists working with the images, noticed

Images of (left to right) Saturn's D, C, B, A, and F
rings, taken in 2007 by the Cassini orbiter.

something totally unexpected: dark shadows like fuzzy spokes of a
wheel, which rotate. No one had seen anything before in the rings that
was not circularly symmetric. Careful analysis of the radii of the rings
revealed another puzzle: one ring is not quite circular.

Voyager 2, which had launched before *Voyager 1* but was moving
more slowly to allow it to continue to Uranus and Neptune, confirmed
these sightings when it passed by Saturn nine months later. As more
and more information flooded in, new puzzles appeared. There are
rings that appeared to be braided, rings with strange kinks, and incomplete rings consisting of several separate arcs with gaps between them.
Previously undetected moons of Saturn were spotted within the rings.
Before the *Voyager* encounters, earthbound astronomers had detected
nine of Saturn's moons. Soon the number rose to more than 30. Today
it's 62, plus a hundred or more tiny moonlets residing in the rings. Of
these satellites, 53 now have official names. The *Cassini* probe, orbiting
Saturn, is providing a stream of data on the planet, its rings, and its
moons.

The moons explain some features of the rings. The main gravitational influence on particles in the rings is Saturn itself. Next in order
of importance are gravitational forces exerted by various moons, especially those nearby. So although ring features seem not to be associated
with resonances with the *main* satellites, we might expect them to be
associated with smaller, but nearer, satellites. This mathematical prediction is borne out spectacularly in the fine structure of the outermost
region of the A ring. Virtually every feature occurs at a distance corresponding to a resonance with the moons Pandora and Prometheus,
which lie on either side of the nearby F ring – a relationship to which we
will shortly return. The relevant resonances, for mathematical reasons,
involve two consecutive integers – for example, 28:27.

The diagram shows the outer edge of the A ring, and the slanting

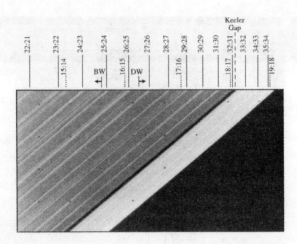

Outer part of the A ring, showing features associated with resonances with Pandora (dotted lines) and Prometheus (solid lines). The grid of dots is imposed by the imaging process.

white lines are regions where the particle density is greater than average. The vertical lines label these orbits with the corresponding resonances: dotted lines for Pandora, solid for Prometheus. Essentially all of the prominent lines correspond to resonant orbits. Also shown in the picture are the locations of a spiral bending wave (BW) and a spiral density wave (DW) that correspond to an 8:5 resonance with another moon, Mimas.

✦

The F ring is very narrow, which is puzzling because narrow rings, if left to their own devices, are unstable and will slowly broaden. The current explanation involves Pandora and Prometheus, but some features are still unsatisfactory.

This issue first showed up in connection with a different planet, Uranus. Until recently, Saturn was the only planet in the solar system (or anywhere else, for that matter) known to have a ring system. But in 1977 James Elliot, Edward Dunham, and Jessica Mink were making observations with the Kuiper Airborne Observatory, a transport aircraft equipped with a telescope and other apparatus. Their intention was to find out more about Uranus's atmosphere.

As a planet moves in orbit, it occasionally passes in front of a star, cutting off some of its light, an event known as an occultation. By measuring the apparent light output of the star as it dims and brightens, astronomers can gain information about the planet's atmosphere by measuring the light curve – how the amount of light (of various wavelengths) changes. In 1977 there was an occultation of the star SAO 158687 by Uranus, and that's what Elliot, Dunham, and Mink were planning to observe. The technique provides information not just about the atmosphere, but about anything that orbits the planet – if it happens to obscure the star. Their light curve showed a series of five tiny blips before the main event, when the star became much dimmer, and a matching series of blips after Uranus had passed across it. Such a blip might just be caused by a tiny moon, but it would have to be in exactly the right place at the right time – twice. A ring, on the other hand, will sweep across the star, so no coincidence is required to affect the light curve. Therefore the most reasonable interpretation of the data was: Uranus has five very thin, faint rings.

When the *Voyagers* encountered Uranus, they confirmed this theory by observing Uranus's rings directly. (Thirteen rings are now known.) They also revealed that the rings are no more than 10 kilometres wide. This seems remarkably narrow, because, as already remarked, a narrow ring should be unstable, slowly broadening as time passes. By understanding the mechanisms that lead to this broadening, we can estimate the likely lifetime of a narrow ring. It turns out that Uranus's rings should last no more than 2500 years. Possibly the rings formed less than 2500 years ago, but it seems unlikely that nine rings should all form so closely spaced in time. The alternative is that some other factor stabilises the rings and prevents them from broadening. In 1979 Peter Goldreich and Scott Tremaine[1] proposed a remarkable mechanism to achieve just that: shepherd moons.

Imagine that the narrow ring in question happens to be just inside the orbit of a small moon. By Kepler's third law, the moon revolves round the planet slightly more slowly than the outer edge of the ring does. Calculations show that this causes the elliptical orbit of a ring particle to become slightly less eccentric – fatter – so its maximum distance from the planet decreases slightly. It looks like the moon is repelling the ring, but actually the effect is the result of gravitational forces that slow the ring particles down.

All very well, but such a moon also disrupts the rest of the ring, especially its inner edge. Solution: add another moon, orbiting just inside the ring. This has a similar effect on the inner edge, but now the moon revolves faster than the ring, so it tends to speed the ring particles up. They therefore move away from the planet, and again it looks as though the moon is repelling the ring.

If a thin ring is sandwiched between two small moons, these effects combine to keep it squashed between their orbits. This cancels out any other tendency that would otherwise cause it to broaden. Such moons are known as shepherd moons, because they keep the ring on track, much as a shepherd controls his flock of sheep. 'Sheepdog moons' might have been a better simile, but the verb 'to shepherd' describes what the moons do. A more detailed analysis shows that the portion of ring trailing behind the inner moon and ahead of the outer will acquire ripples, but these die down as a consequence of collisions between the ring particles.

When *Voyager 2* reached Uranus, one of its images showed that Uranus's ε ring sits neatly between the orbits of two moons, Ophelia and Cordelia. (The rings of Uranus are labelled by Greek lowercase letters, and ε is the letter epsilon.) Goldreich and Tremaine's theory was vindicated. Resonances are involved too. The outer edge of Uranus's ε ring corresponds to a 14:13 resonance with Ophelia, and the inner edge corresponds to a 24:25 resonance with Cordelia.

Saturn's F ring is similarly situated between the orbits of Pandora and Prometheus, and it's believed that this is a second instance of shepherd moons. However, there are complications, because the F ring is surprisingly dynamic. *Voyager 1* images of November 1980 show the F ring with clumps, kinks, and an apparently braided segment. When *Voyager 2* passed by in August 1981 hardly any of these features could be seen, just one braid-like section. It's now thought that the other features disappeared between the two encounters, implying that changes in the form of the F ring can occur within a few months.

The suggestion is that these transient dynamic effects are also caused by Pandora and Prometheus. Waves generated by close approaches of the moons don't die away, so traces of them remain the next time the moon passes by. This makes the dynamics of the ring more complex, and it also means that the tidy explanation of a narrow ring held in place by shepherd moons is too simplistic. Moreover, the orbit of Prometheus is chaotic because of a 121:118 resonance with Pandora, but only Prometheus contributes to the confinement of the F ring. So although the shepherd moon theory provides some insight into the narrowness of the F ring, it's not the whole story.

As further evidence, the inner and outer edges of the F ring don't correspond to resonances. In fact, the strongest resonances near the F ring are associated with two other moons altogether, Janus and Epimetheus. These moons have their own bizarre behaviour: they are co-orbital. Taken literally, this term ought to mean 'share the same orbit', and in a sense they do. Most of the time, one of them has an orbit a few kilometres larger than the other's. Since the innermost one moves faster, it would collide with the outer moon if they stuck to their elliptical orbits. Instead, they interact, and *swap places* with each other. This happens every four years. That's why I said 'innermost' and 'outer'. Which is which depends on the date.

This kind of switching differs wildly from the tidy ellipses that Kepler envisaged. It happens because ellipses are the natural orbits for *two*-body dynamics. When a third body enters the picture, orbits take on new shapes. Here, the effect of the third body is mostly small enough to ignore, so each moon pretty much follows an ellipse, as if the other didn't exist. But when they get close together, that approximation fails. They interact, and in this case they swing round each other, so each moon moves into the other moon's previous orbit. There is a sense in which the true orbit of each moon can be described as one ellipse alternating with the other, with short transitional paths between the two. Both moons follow such an orbit, based on the same two ellipses. They just make the transition in opposite directions at the same time.

✦

Saturn's rings have been known since the time of Galileo, even though he wasn't sure what they were. Uranus's rings came to human attention in 1979. We now know that Jupiter and Neptune also have very faint ring systems. Saturn's moon Rhea may have its own very tenuous ring system.

Moreover, Douglas Hamilton and Michael Skrutskie discovered in

2009 that Saturn has an absolutely gigantic, but very faint, ring, far larger than the ones Galileo and the *Voyagers* saw. They missed it, in part, because it's visible only in infrared light. Its inner edge is about 6 million kilometres from the planet, and the outer edge is 18 million kilometres distant. The moon Phoebe orbits inside it, and may well be responsible for it. It's very tenuous, made from ice and dust, and it may help to solve a long-standing mystery: Iapetus's dark side. One half of the moon Iapetus is brighter than the other, an observation that has puzzled astronomers since about 1700 when Cassini first noticed it. The suggestion is that Iapetus is sweeping up dark material from the giant ring.

In 2015 Matthew Kenworthy and Eric Mamajek announced[2] that a distant exoplanet, orbiting the star J1407, has a system of rings that makes Saturn's pale into insignificance, even taking the newest one into account. The discovery, like the rings of Uranus, is based on fluctuations observed in a light curve – the main way to locate exoplanets (see Chapter 13). As the planet crosses the star (transits) the star's light dims. In this instance, it dimmed repeatedly, over a two-month period, but each dimming event was fairly rapid. The inference is that some exoplanet with numerous rings must have been passing across the path from the star to the Earth. The best ring model has 37 rings and extends out to a radius of 0·6 AU (90 million km). The exoplanet itself hasn't yet been detected, but it's thought to be about 10–40 times as massive as Jupiter. A clear gap in the ring system is most readily explained by the presence of an exomoon, whose size can also be estimated.

In 2014 another ring system was discovered in the solar system in an improbable place: around (10199) Chariklo, a type of small body known as a centaur.[3] This one orbits between Saturn and Uranus, and it's the largest known centaur. Its rings showed up as two slight dips of brightness in a series of observations in which Chariklo obscured (jargon: occulted) various stars. The relative positions of these dips are all close to the same ellipse, with Chariklo near the centre, suggesting two nearby rings in fairly circular orbits, whose plane is being viewed at an angle. One has radius 391 km and is about 7 km across; then there's a 9 km gap to the second, at radius 405 km.

Since ring systems occur repeatedly, they can't just be an accident. So how do ring systems form? There are three main theories. They may have formed when the original gas disc coalesced to create the planet;

they could be relics of a moon that has been broken by a collision; they could be remains of a moon that got closer than the Roche limit, at which tidal forces exceed the strength of the rock, and broke up.

Catching a ring system during its formation is unlikely, although Kenworthy and Mamajek's discovery shows that it's possible, but the best that it can give us is a snapshot. Observing the entire process would take hundreds of lifetimes. What we can do, though, is to analyse hypothetical scenarios mathematically, make predictions, and compare them with observations. It's like fossil-hunting in the heavens. Each 'fossil' provides evidence for what happened in the past, but you need a hypothesis to interpret the evidence, and you need mathematical simulations or inferences or, better still, theorems to understand the consequences of that hypothesis.

Cosimo's Stars

Since it is up to me, the first discoverer, to name these new planets, I wish, in imitation of the great sages who placed the most excellent heroes of that age among the stars, to inscribe these with the name of the Most Serene Grand Duke [Cosimo II de' Medici, Grand Duke of Tuscany].

Galileo Galilei, *Sidereus Nuncius*

WHEN GALILEO FIRST OBSERVED JUPITER through his new telescope, he noticed four tiny specks of light in orbit around the planet, so Jupiter had moons of its own. This was direct evidence that the geocentric theory must be wrong. He sketched their arrangements in his notebook. More detailed observations can be strung together, so we can plot the paths along which the specks seem to move. When we do, we obtain beautiful sine curves. The natural way to generate a sine curve is to observe uniform circular motion sideways on. So Galileo inferred that Cosimo's stars go round Jupiter in circles, in the plane of the ecliptic.

Improved telescopes revealed that most planets in the solar system have moons. Mercury and Venus are the sole exceptions. We've got 1, Mars has 2, Jupiter has at least 67, Saturn at least 62 plus hundreds of moonlets, Uranus 27, and Neptune 14. Pluto has 5. The satellites range from small irregular rocks to nearly spherical ellipsoids big enough to be small planets. Their surfaces can be mainly rock, ice, sulphur, or frozen methane.

Mars's tiny moons Phobos and Deimos race across the Martian sky, and Phobos is so close that it moves in the opposite direction to Deimos. Both bodies are irregular, and are probably captured asteroids

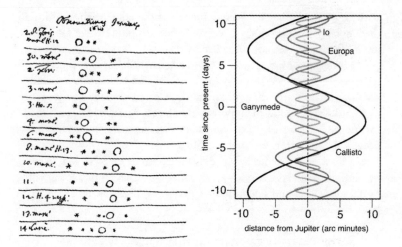

Left: Galileo's records of the moons. Right: Positions of Jupiter's
moons as seen from Earth, forming sine curves.

– or, possibly, a captured asteroid with a duck shape like comet 67P,
recently shown to be two bodies that came gently together and stuck.
If so, the one that Mars captured came unstuck again because of the
planet's gravity; Phobos is one piece and Deimos the other.

Some moons seem totally dead; others are active. Saturn's moon
Enceladus produces towering ice geysers 500 kilometres high. Jupiter's
moon Io has a sulphurous surface and at least two active volcanos, Loki
and Pele, spewing out sulphur compounds. There must be huge subsurface reservoirs of liquid sulphur, and the energy to heat it probably
comes from gravitational squeezing by Jupiter. Saturn's moon Titan
has a methane atmosphere, much denser than it ought to be. Neptune's
moon Triton is going round the planet the wrong way, indicating that it
was captured. It's spiralling slowly inwards, and 3·6 billion years from
now it will break up when it hits the Roche limit, the distance inside
which moons come to pieces under gravitational stress.

The moons of the larger planets often exhibit resonances. For
example, Europa has twice the period of Io, and Ganymede has twice
the period of Europa, hence four times that of Io. Resonances come
from the dynamics of bodies obeying Newton's law of gravity. Planets

with ring systems slowly accrete moons at the edge of the rings, and then 'spit them out' one by one, like water dripping from a tap. There are mathematical regularities in this process.

Various lines of evidence, some mathematical, indicate that several icy moons have underground oceans, melted by tidal forces. At least one contains more water than all Earth's oceans combined. The presence of liquid water makes them potential habitats for simple Earthlike life forms, see Chapter 13. And Titan's unusual chemistry might make it a potential habitat for un-Earthlike life forms.

At least one asteroid has a very tiny moon of its own: Ida, orbited by diminutive Dactyl. Moons are fascinating, a playground for gravitational modelling and scientific speculation of all kinds. And it all goes back to Galileo and Cosimo's stars.

Asteroid Ida (left) and its moon Dactyl (right).

✦

In 1612, when Galileo had determined the orbital periods of Cosimo's stars, he suggested that sufficiently accurate tables of their motion would provide a clock in the sky, solving the longitude problem in navigation. At that time navigators could estimate their latitude by observing the Sun (though accurate instruments like the sextant were far in the future), but longitude relied on dead reckoning – educated guesswork. The main practical issue was to make observations from the deck of a ship as it tossed on the waves, and he worked on two devices for stabilising a telescope. The method was used on land but not at sea. John Harrison famously solved the longitude problem with his series of very accurate chronometers, eventually being awarded prize money in 1773.

Jupiter's moons presented astronomers with a celestial laboratory, allowing them to observe systems of several bodies. They tabulated their movements and attempted to explain and predict them theoretically. One way to obtain precise measurements is to observe a transit of a moon across the face of the planet, because the start and end of a transit are well-defined events. Eclipses, where the moon goes behind the planet, are similarly well defined. Giovanni Hodierna said as much in 1656, and a decade or so later Cassini began a lengthy series of systematic observations, noting other coincidence such as conjunctions, in which two of the moons appear to be aligned. To his surprise, the transit times didn't seem consistent with the moons moving in regular, repetitive orbits.

The Danish astronomer Ole Rømer took up Galileo's suggestion about longitude, and in 1671 he and Jean Picard observed 140 eclipses of Io from Uraniborg, near Copenhagen, while Cassini did the same from Paris. Comparing the timings, they calculated the difference between the longitudes of those two locations. Cassini had already noticed some peculiarities in the observations, and wondered whether they resulted from light having a finite speed. Rømer combined all the observations, and discovered that the time between successive eclipses became shorter when Earth was closer to Jupiter, and longer when it was further away. In 1676 he informed the Academy of Sciences of the reason: 'Light seems to take about ten to eleven minutes [to cross] a distance equal to the radius of the terrestrial orbit.' This figure relied on some careful geometry, but the observations were inaccurate; the modern value is 8 minutes 12 seconds. Rømer never published his results as a formal paper, but the lecture was summarised – badly – by an unknown reporter. Scientists didn't accept that light has a finite speed until 1727.

Despite the irregularities, Cassini never observed a triple conjunction of the inner moons, Io, Europa, and Ganymede – all three of them aligning simultaneously – so something must prevent this. Their orbital periods are roughly in the ratio 1:2:4, and in 1743 Pehr Wargentin,

Director of the Stockholm Observatory, showed that this relationship becomes astonishingly accurate if it's reinterpreted correctly. Measuring their positions as angles relative to a fixed radius, he discovered a remarkable relation:

angle for Io − 3 × angle for Europa + 2 × angle for Ganymede = 180°

According to his observations, this equation holds almost exactly over long periods of time, *despite* the irregularities in the three moons' orbits. A triple conjunction requires the three angles to be equal, but if they are, the left-hand side of this equation is 0°, not 180°. So a triple conjunction is impossible as long as the relationship holds good. Wargentin stated that it wouldn't happen for at least 1·3 million years.

The equation also implies a specific pattern to conjunctions of these moons, which occur in a repeating cycle:

Europa with Ganymede
Io with Ganymede
Io with Europa
Io with Ganymede
Io with Europa
Io with Ganymede

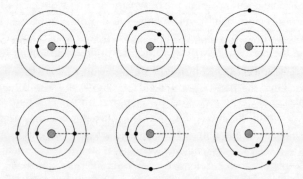

Successive conjunctions of the three innermost moons of
Jupiter: Io, Europa, and Ganymede (reading outwards).

Laplace decided that Wargentin's formula couldn't be coincidence, so there must be a dynamical reason. In 1784 he deduced the formula from Newton's law of gravitation. His calculations imply that over long periods the combination of angles concerned does not remain exactly at 180°; instead, it librates – oscillates slowly either side of that value – by less than 1°. This is small enough to prevent a triple conjunction. He predicted the period of this oscillation to be 2270 days. The observed figure today is 2071 days, not bad. In his honour the relationship between the three angles is called the Laplace resonance. His success was a significant confirmation of Newton's law.

We now know why the transit times are irregular. Jupiter's gravity causes the approximately elliptical orbits of its moons to precess (like Mercury's orbit round the Sun), so that the position of perijove – the closest approach to Jupiter – changes quite fast. In the Laplace resonance formula these precessions cancel out, but they have a strong effect on individual transits.

Any similar relationship is also called a Laplace resonance. The star Gliese 876 has a system of exoplanets, the first being found in 1998. Four are now known, and three of them – Gliese 876c, Gliese 876b, and Gliese 876e – have orbital periods of 30·008, 61·116, and 124·26 days, suspiciously close to ratios 1:2:4. In 2010 Eugenio Rivera and colleagues[1] showed that in this case the relationship is:

angle for 876c − 3 × angle for 876b + 2 × angle for 876e = 0°

but the sum librates around 0° by as much as 40°, a far bigger oscillation. Now triple conjunctions are possible, and near triple conjunctions occur once for every revolution of the outermost planet. Simulations indicate that the oscillation around 0° should be chaotic, with an approximate period of about ten years.

Three of Pluto's moons – Nix, Styx, and Hydra – exhibit a Laplace-like resonance, but this one has mean period ratios 18:22:33 and mean orbital ratios 11:9:6. The equation is now:

3 × angle for Styx − 5 × angle for Nix + 2 × angle for Hydra = 180°

Triple conjunctions are impossible, by the same reasoning as for the Jovian moons. There are five Styx/Hydra conjunctions and 3 Nix/Hydra conjunctions for every two Styx/Nix conjunctions.

✦

Europa, Ganymede, and Callisto all have icy surfaces. Several lines of evidence indicate that all three of them have oceans of liquid water beneath the ice. The first moon to be suspected of hosting such an ocean was Europa. Some heat source is needed to melt the ice. Tidal forces from Jupiter squeeze Europa repeatedly, but resonances with Io and Ganymede prevent it escaping by changing orbit. The squeezing heats the core of the moon, and calculations suggest that the amount of heat is enough to melt much of the ice. Since the surface is solid ice, the liquid water must be deeper down, probably forming a thick spherical shell.

As further support, the surface is extensively cracked, with few signs of craters. The most likely explanation is that the ice forms a thick layer floating on the ocean. Jupiter's strong magnetic field induces a weaker one in Europa, and when the *Galileo* orbiter measured Europa's magnetic field, mathematical analysis suggested that a substantial mass of conducting material must lie under Europa's ice. The most plausible substance, given the data, is salty water.

Europa's surface has a number of regions of 'chaos terrain' where the ice is very irregular and jumbled. One such is Conamara Chaos, which appears to be formed from innumerable ice rafts that have been broken and moved. Others are Arran, Murias, Narbeth, and Rathmore Chaos. Similar formations occur on Earth in pack ice, floating on the sea, when a thaw sets in. In 2011 a team led by Britney Schmidt explained that chaotic terrain forms when ice sheets, lying above lens-shaped lakes of liquid water, collapse. These lakes are nearer the surface than the ocean itself, perhaps only three kilometres underground.[2] A depression of this kind called Thera Macula has an underlying lake with as much water as the North American Great Lakes.

Europa's lenticular lakes are closer to the surface than the main ocean. The best estimates right now are that aside from such lakes, the outer layer of ice is about 10–30 km thick, and the ocean is 100 km deep. If so, Europa's ocean has twice the volume of all of Earth's oceans combined.

Based on similar evidence, Ganymede and Callisto also have subsurface oceans. Ganymede's outer layer of ice is thicker, about 150 km, and the ocean beneath is again about 100 km deep. Callisto's ocean probably lies the same distance under the ice, with an ocean 50–200 km deep. All of these figures are speculative, and differences in the chemistry, such as the presence of ammonia, would change them significantly.

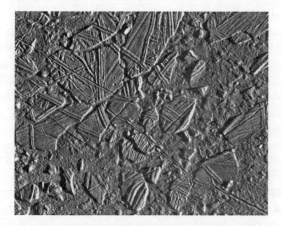

Conamara Chaos on Europa.

Enceladus, a moon of Saturn, is very cold, with a mean surface temperature of 75 K (about *minus* 200°C). You'd expect it not to show much activity, and so did astronomers, until *Cassini* discovered that it emits huge geysers of ice particles, water vapour, and sodium chloride, hundreds of kilometres high. Some of this material escapes altogether, and is believed to be the main source of Saturn's E ring, which contains 6% sodium chloride. The rest falls back to the surface. The most plausible explanation, a salty underground ocean, was confirmed in 2015 by a mathematical analysis of seven years' worth of data on tiny wobbles in the moon's orientation (jargon: libration), measured by observing accurate positions of its craters.[3] The moon wobbles through an angle of 0·12 degrees. This is too large to be consistent with a rigid connection between Enceladus's core and its icy surface, and it indicates a global ocean rather than a more limited polar sea. The ice above is probably 30–40 km thick, and the ocean is 10 km deep – more than the average for Earth's oceans.

✦

Seven of Saturn's moons orbit just beyond the edge of the planet's outer main ring, the A ring. They're very small and their density is extremely low, suggesting they have internal voids. Several are shaped like flying saucers, and some have smooth patchy surfaces. They are Pan, Daphnis, Atlas, Prometheus, Pandora, Janus, and Epimetheus.

In 2010 Sébastian Charnoz, Julien Salmon, and Aurélien Crida[4] analysed how the ring might evolve, along with hypothetical 'test bodies' at its edge, concluding that these moons have been spat out by the rings as material passes outside the Roche limit. The Roche limit is usually defined to be the distance inside which moons come to pieces under gravitational stress, but conversely it's also the distance outside which rings become unstable, unless stabilised by other mechanisms such as shepherd moons. Saturn's Roche limit (140,000 ± 2000 km) is just outside the edge of the A ring (136,775 km). Pan and Daphnis are just inside the Roche limit, the other five are just outside.

Astronomers had long suspected that there must be a connection between the rings and these moons, because their radial distances are so close together. The A ring has a very sharp boundary, created by a 7:6 resonance with Janus, which prevents most of the ring material moving further out. This resonance is temporary; the rings 'push' Janus further out while themselves initially moving inwards a little to conserve angular momentum. As Janus continues to move out, the rings can spread outwards again, passing the Roche limit.

The analysis supports this view, showing that some ring material can temporarily be pushed outside the Roche limit by viscous spreading – much as a blob of syrup on the kitchen table will slowly spread out and become thinner. Their method combines an analytic model of the test bodies with a numerical fluid-dynamics model of the rings. Continued viscous spreading causes the rings to spit out a succession of tiny moonlets, whose orbits resemble reality quite closely. Calculations indicate that these moonlets are aggregates of ice particles from the rings, loosely held together by their own gravity, explaining their low density and curious shapes.

The results also shed light on a long-standing question: the age of the rings. One theory is that the rings formed from the collapsing solar nebula at much the same time as Saturn did. However, a moonlet such as Janus should take no more than a hundred million years to drift outwards from the A ring to its current orbit, suggesting an alternative theory: both the rings and these moonlets appeared together when a larger moon passed inside the Roche limit and broke up, some tens of millions of years ago. The simulations reduce this period to between 1 and 10 million years: the authors say, 'Saturn's rings, like a mini protoplanetary disk, may be the last place where accretion was recently active in the solar system, some 10^6–10^7 years ago.'

Off on a Comet

> It may be said with perfect truth that a fisherman standing on the sun's surface, holding a rod long enough, could fling his line in no direction without hooking plenty of comets.
>
> Jules Verne, *Off on a Comet!*

'WHEN BEGGARS DIE, THERE are no comets seen; the heavens themselves blaze forth the death of princes.' So says Calpurnia in Act 2, Scene 2 of Shakespeare's *Julius Caesar*, prophesying Caesar's demise. Of the five references to comets in Shakespeare, three reflect the ancient belief that comets are harbingers of disaster.

These strange and puzzling objects appear unexpectedly in the night sky trailing a bright curved tail, move slowly across the background of stars, and disappear again. They are unheralded interlopers that don't seem to fit the normal patterns of celestial events. Reasonable enough, then, in times when none knew better, and when priests and shamans were always seeking to bolster their influence, to interpret a comet as a messenger from the gods. The common assumption was that the message was ominous. There were enough natural disasters around that if that's what you wanted to believe, it wasn't hard to find convincing confirmation. Comet McNaught, which appeared in 2007, was the brightest for 40 years. Clearly it heralded the 2007–8 financial crisis. See? Anyone can do it.

Priests claimed to know what comets were for, but neither they nor philosophers knew where they were located. Were they celestial bodies, like the stars and planets? Or were they meteorological phenomena, like clouds? The *looked* a bit like clouds; they were fuzzy, not crisp like

Great Comet of 1577 over Prague. Engraving by Jiri Daschitzky.

stars and planets. But they moved more like planets, except for those sudden appearances and disappearances. Ultimately, the debate was settled by scientific evidence. When the astronomer Tycho Brahe used precision measurements to estimate the distance to the Great Comet of 1577, he demonstrated that it was further away than the Moon. Since clouds hide the Moon but not the other way round, comets are of celestial habitat.

✦

By 1705 Edmond Halley had gone further, showing that at least one comet is a regular visitor to the night skies. Comets are like planets: they orbit the Sun. They seem to disappear when they get too far away to see, and reappear when they get close enough again. Why do they grow tails, and lose them again? Halley wasn't sure, but it was related to their proximity to the Sun.

Halley's insight into comets was one of the first big discoveries in astronomy to be deduced from the mathematical patterns discovered by Kepler and reinterpreted more generally by Newton. Since planets move in ellipses, Halley reasoned, why not comets too? If so, their motion would be periodic, and the same comet would return repeatedly to Earthly skies at equally spaced times. Newton's law of gravity modified this statement a little: the motion would be *almost* periodic, but the gravitational attraction of other planets, especially the giants Jupiter and Saturn, would accelerate or retard the comet's return.

To test this theory, Halley delved into arcane records of comet sightings. Before Galileo's invention of the telescope, only comets visible to the naked eye could be seen. A few were unusually bright, with an impressive tail. Petrus Apianus saw one in 1531; Kepler had observed another in 1607; a similar comet had appeared during Halley's lifetime, in 1682. The gaps between these dates are 76 and 75 years. Might all three sightings be of the *same* body? Halley was convinced they were, and he predicted that the comet would return in 1758.

He was right, just. On Christmas day of that year the German amateur astronomer Johann Palitzsch saw a faint smudge in the sky, which soon developed the characteristic tail. By then, three French mathematicians, Alexis Clairaut, Joseph Lalande, and Nicole-Reine Lepaute, had carried out more accurate calculations, amending the predicted date for the comet's closest approach to the Sun to 13 April. The actual date was a month earlier, so perturbations by Jupiter and Saturn had delayed the comet by 618 days.

Halley died before his prediction could be verified. What we now call Halley's comet (named after him in 1759) was the first body other than a planet to be shown to orbit the Sun. By comparing ancient records to modern computations of its past orbit, Halley's comet can be traced back to 240 BC, when it was seen in China. Its next appearance in 164 BC was noted on a Babylonian clay tablet. The Chinese saw it again in 87 BC, 12 BC, 66 AD, 141 AD, ... and so on. Halley's forecast of the comet's eventual return was one of the earliest truly novel astronomical predictions to be based on a mathematical theory of celestial dynamics.

✦

Comets aren't just an abstruse astronomical puzzle. In the introduction I mentioned a far-reaching theory involving comets: for the past few decades they've been the preferred explanation of how the Earth got its oceans. Comets are mainly composed of ice; the tail forms when the comet gets close enough to the sun for the ice to 'sublimate', that is, turn directly from solid to vapour. There's convincing circumstantial evidence that lots of comets collided with the early Earth, in which case the ice would melt and collect to form oceans. Water would also be incorporated into the molecular structure of crustal rocks, which actually contain quite a lot of it.

Earth's water is vital to the planet's life forms, so understanding comets has the potential to tell us something important about ourselves. Alexander Pope's 1734 poem *An Essay on Man* includes the memorable line 'the proper study of mankind is Man'. However, without going into the poem's spiritual and ethical intentions, any study of humanity should also involve the *context* for human beings, not just the beings themselves. That context is the entire universe – so, Pope's dictum notwithstanding, the proper study of mankind is *everything*.

Today, astronomers have catalogued 5253 comets. There are two main types: long-period comets with periods 200 years or greater, whose orbits extend beyond the outer reaches of the solar system, and short-period comets that stay closer to the Sun and often have rounder, though still elliptical, orbits. Halley's comet, with its 75-year period, is a short-period comet. A few comets have hyperbolic orbits: we've already encountered the hyperbola, another conic section familiar to ancient Greek geometers, on page 17. Unlike an ellipse, it doesn't close up. Bodies following such an orbit appear from a vast distance, swing past the Sun, and if they manage not to collide with it they head back out into space, never to be seen again.

A hyperbolic shape suggests these comets fall Sunwards from interstellar space, but astronomers now think that most of them, perhaps all, originally followed very distant closed orbits before being perturbed by Jupiter. The distinction between ellipses and hyperbolas involves the energy of the body. Below a critical value of the energy, the orbit is a closed ellipse. Above that value, it's a hyperbola. At that value, it's a parabola. A comet in a very large elliptical orbit, perturbed by Jupiter, gains extra energy, which can be enough to tip it past the critical value. A close encounter with an outer planet can add more energy through the slingshot effect: the comet steals some of the planet's energy, but the planet is so massive that it doesn't notice. In this manner, the orbit can become a hyperbola.

A parabolic orbit is unlikely because it's poised at the critical energy value. But, for just that reason, a parabola was often used as a first step towards computing a comet's orbital elements. A parabola is close to both an ellipse and a hyperbola.

✦

This brings us back to the short-period comet that hit the headlines, named 67P/Churyumov–Gerasimenko after its discoverers Klim Churyumov and Svetlana Gerasimenko. It orbits the Sun every six and a half years. 67P's hitherto commonplace cometary existence, pottering round the Sun and expelling jets of heated water vapour when it got too close, came to the attention of astronomers, and the *Rosetta* space-craft was sent to rendezvous with it. As *Rosetta* approached its goal, 67P was revealed as a cosmic rubber duck: two round lumps joined by a narrow neck. At first no one was sure whether this shape arose from two rounded bodies that came together very slowly, or a single body that eroded in the neck region.

Late in 2015 this issue was resolved by an ingenious application of mathematics to detailed images of the comet. At first sight 67P's terrain seems jumbled and irregular, with jagged cliffs and flat depres-sions assembled at random, but its surface detail provides clues to its origins. Imagine taking an onion, slicing off random pieces, and hacking chunks out of it. Thin slices parallel to the surface would leave flat spots, deeper gouges would show a stack of separate layers. The comet's flat depressions are akin to the slices, and its cliffs and other regions often show layered strata of ice. Series of layers can be seen at the top and centre right of the image on page 2, for example, and many flat regions are visible.

Astronomers think that when comets first appeared in the early solar system they grew by accretion, so that layer upon layer of ice was gradually added, much like the layers of an onion. So we can ask whether the geological formations visible in images of 67P are consis-tent with this theory, and if so, we can use the geology to reconstruct the comet's history.

Matteo Massironi and coworkers carried out this task in 2015.[1] Their results offer strong support for the theory that the duck shape was created by a gentle collision. The basic idea is that the comet's history can be deduced from the geometry of its ice layers. Eyeballing the images, the two-body theory looks a better bet, but Massironi's team carried out a careful mathematical analysis using three-dimen-sional geometry, statistics, and mathematical models of the comet's gravitational field. Starting from a mathematical representation of the

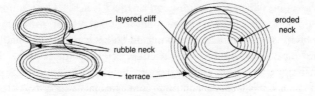

Two competing scenarios for the structure of 67P. Left:
Collision theory. Right: Erosion theory.

observed shape of the comet's surface, the team first worked out the
positions and orientations of 103 planes, each providing the best fit to a
geological feature associated with the observed layers, such as a terrace
(flat region) or cuesta (a type of ridge). They found that these planes fit
together consistently around each lobe, but not at the neck where the
lobes join. This indicates that each lobe acquired onion-like layers as it
grew, before they came together and stuck.

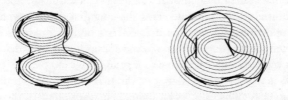

Schematic illustration of best-fitting planes to terraces and cuestas. Left: Collision
theory. Right: Erosion theory. The actual calculation was performed in three
dimensions using a statistical measure for the best fit, and used 103 planes.

When the layers form, they're roughly perpendicular to the local
direction of gravity – a technical way to say that the additional material
falls *downwards*. So for further confirmation, the team computed the
gravitational field of the comet under each of the two hypotheses, and
used statistical methods to show that the layers fit the collision model
better.

Despite being made mainly of ice, 67P is black as midnight and
pockmarked with thousands of rocks. *Philae* made a difficult, and as

it turned out, temporary, landing on the duck's head. The landing did not go as intended. *Philae*'s equipment included a small rocket motor, spikes with screw threads, harpoons, and a solar panel. The plan was to make a gentle landing, fire the rocket to keep the lander pressed against the comet's surface, harpoon the comet to hold it in place when the rockets were turned off, screw the spikes into the comet to make sure it stayed there, and then use the solar panel to harvest energy from sunlight. Men, mice, plans, best-laid ... the rocket failed to fire, the harpoons failed to stick, the screws failed to bite, and as a result the solar panel ended up in deep shadow with hardly any sunlight to harvest.

Despite its proverbial 'perfect three-point landing – two knees and a nose', *Philae* attained almost all of its scientific objectives, sending back vital data. It was hoped that it might add more as the comet got closer to the Sun, the light got stronger, and the probe woke up from its electronic sleep. *Philae* did briefly renew contact with the ESA, but communication was lost again, probably because the increasing activity of the comet damaged it. Before it ran out of power, *Philae* confirmed that the comet's surface is ice with a coating of black dust. As already mentioned, it also sent measurements showing that the ice contains a greater proportion of deuterium than Earth's oceans, casting serious doubt on the theory that the water in the oceans was mainly delivered by comets when the solar system was forming.

Ingenious work using the data that did make it home has provided further useful information. For example, mathematical analysis of how *Philae*'s landing struts compressed shows that in places 67P has a hard crust, but elsewhere the surface is softer. Images made by *Rosetta* include three marks where the lander first hit the comet, deep enough to show that the material there is relatively soft. *Philae*'s on-board hammer was unable to penetrate the ice where it came to rest, so there the ground is hard. On the other hand, the bulk of 67P is very porous: three quarters of its interior is empty space.

Philae sent back some intriguing chemistry, too: several simple organic (this means carbon-based, and is not indicative of life) compounds, and a more complex one, polyoxymethylene, probably created from the simpler molecule formaldehyde by the action of sunlight. Astronomers were startled by one of *Rosetta*'s chemical discoveries: quite a lot of oxygen molecules in the gas cloud surrounding the comet.[2] They were so surprised that at first they assumed they'd made a mistake. In conventional theories of the origin of the solar system, the oxygen would have been heated, making it react with other elements to form compounds such as carbon dioxide, so it would no longer be around as pure oxygen. The early solar system must have been less violent than previously thought, allowing grains of solid oxygen to build up slowly and avoid forming compounds.

That doesn't conflict with the more dramatic events that are thought to have occurred during the formation of the solar system, such as migrating planets and colliding planetesimals, but it suggests that such events must have been relatively rare, punctuating a background of slow, gentle growth.

✦

Where do comets come from?

Long-period comets can't hang around indefinitely in their present orbits. As they pass through the solar system, there's a risk of a collision, or a close encounter that hurls them off into space, never to return. The chance may be small, but over millions of orbits the odds against avoiding such disasters mount up. Moreover, comets decay, losing mass every time they round the Sun and stream off sublimating ice. Hang around too long, and they melt away.

In 1932 Ernst Öpik suggested a way out: there must be a huge reservoir of icy planetesimals in the outer reaches of the solar system, which replenishes the supply of comets. Jan Oort had the same idea independently in 1950. From time to time one of these icy bodies is dislodged, perhaps by near misses with another one, or just by chaotic gravitational perturbations. It then changes its orbit, falling in towards the Sun, warms up, and the characteristic coma and tail are born. Oort investigated this mechanism in considerable mathematical detail, and in his honour we now name the source the Oort cloud. (As explained earlier for asteroids, the name should not be taken literally. It's a very sparse cloud.)

The Oort cloud is thought to occupy a vast region round the Sun between about 5000 AU out to 50,000 AU (0·03 to 0·79 light years). The inner cloud, out to 20,000 AU, is a torus roughly aligned with the ecliptic; the outer halo is a spherical shell. There are trillions of bodies a kilometre or more across in the outer halo, and the inner cloud contains a hundred times as many. The total mass of the Oort cloud is about five times that of the Earth. This structure has not been observed: it's deduced from theoretical calculations.

Simulations and other evidence suggest that the Oort cloud came into existence when the local protoplanetary disc began to collapse and form the solar system. We've discussed evidence that the resulting planetesimals were originally closer to the Sun, and were then hurled into the outermost regions by the giant planets. The Oort cloud could be a remnant of the early solar system formed from leftover debris. Alternatively, it may be the result of a competition between the Sun and neighbouring stars to attract material that was always that far out, near the borderline where the two stars' gravitational fields cancel each other. Or, as proposed in 2010 by Harold Levison and coworkers, the Sun stole debris from the protoplanetary discs of the cluster of 200 or so stars in its vicinity.

If the ejection theory is correct, the initial orbits of the bodies in the Oort cloud were very long, thin ellipses. However, since these bodies mostly stay in the cloud, their orbits must now be much fatter, almost circular. It's thought that the orbits were fattened up by interaction with nearby stars and galactic tides – the overall gravitational effect of the Galaxy.

✦

Short-period comets are different, and are thought to have a different origin: the Kuiper belt and the scattered disc.

When Pluto was discovered, and found to be quite small, many astronomers wondered if it might be another Ceres – the first new body in a huge belt containing thousands. One – not the first – was Kenneth Edgeworth, who suggested in 1943 that when the outer solar system past Neptune condensed from the primal gas cloud, the matter wasn't dense enough to form large planets. He also saw these bodies as a potential source of comets.

In 1951 Gerald Kuiper proposed that a disc of small bodies might have collected in that region early in the formation of the solar system, but he thought (as many then did) that Pluto was about the size of the Earth, so it would have disturbed the disc and scattered its contents far and wide. When it turned out that such a disc still exists, Kuiper received the dubious honour of having an astronomical region named after him because he did *not* predict it.

Several individual bodies were discovered in this region: we've encountered them already as TNOs. What clinched the existence of the Kuiper belt was, once more, comets. In 1980 Julio Fernández carried out a statistical study of short-period comets. There are too many for them all to have come from the Oort cloud. Out of every 600 comets emanating from the Oort cloud, 599 would become long-period comets and only one would be captured by a giant planet and change to a short-period orbit. Perhaps, Fernández said, there's a reservoir of icy bodies between 35 and 50 AU from the Sun. His ideas received strong support from a series of simulations carried out by Martin Duncan, Tom Quinn, and Scott Tremaine in 1988, who also noted that short-period comets stay close to the ecliptic, but long-period ones arrive from almost any direction. The proposal became accepted, with the name 'Kuiper belt'. Some astronomers prefer 'Edgeworth–Kuiper belt' and others assign credit to neither.

The origins of the Kuiper belt are murky. Simulations of the early solar system indicate the scenario mentioned earlier, in which the four giant planets originally formed in a different order (reading outwards from the Sun) from today's, and then migrated, scattering planetesimals to the four winds. Most of the primordial Kuiper belt was flung away, but one body in a hundred remained. Like the inner region of the Oort cloud, the Kuiper belt is a fuzzy torus.

The distribution of matter in the Kuiper belt isn't uniform; like the asteroid belt, it's modified by resonances, in this case with Neptune. There's a Kuiper cliff at about 50 AU, where the number of bodies falls off suddenly. This has not been explained, although Patryk Lyakawa speculates that it might be the result of an undetected large body – a genuine Planet X.

The scattered disc is even more enigmatic and less well known. It overlaps the Kuiper belt slightly, but it extends further, to about 100 AU, and is strongly inclined relative to the ecliptic. Bodies in the scattered disc have highly elliptical orbits and are often diverted into the inner solar system. There they linger for a time as centaurs, before the orbit changes again and they turn into short-period comets. Centaurs are bodies occupying orbits that cross the ecliptic between the orbits of

Jupiter and Neptune; they persist for only a few million years, and there are probably about 45,000 of them more than a kilometre across. The majority of short-period comets probably come from the scattered disc rather than the Kuiper belt.

✦

In 1993 Carolyn and Eugene Shoemaker and David Levy discovered a new comet, later named Shoemaker–Levy 9. Unusually, it had been captured by Jupiter and was in orbit around the giant planet. Analysis of its orbit indicated that the capture had occurred 20–30 years earlier. Shoemaker–Levy 9 was unusual in two ways. It was the only comet observed orbiting a planet, and it seemed to have broken into pieces.

Shoemaker-Levy 9 on 17 May 1994.

The reason emerged from a simulation of its orbit. Calculating backwards, in 1992 the comet must have passed inside Jupiter's Roche limit. Gravity's tidal forces then broke the comet up, creating a string of about 20 fragments. The comet had been captured by Jupiter around 1960–70, and the close encounter had distorted its orbit into a long, thin one.

Simulating the orbit in forward time predicted that on the comet's next fly-by, in July 1994, it would collide with Jupiter. Astronomers had never observed a celestial collision before, so this discovery caused considerable excitement. A collision would stir up Jupiter's atmosphere, making it possible to find out more about its deeper layers, normally hidden by the cloud above. In the event, the impact was even more dramatic than expected, leaving a chain of gigantic scars across the planet that slowly faded, remaining visible for months. Twenty-one impacts in all were sighted; the largest produced 600 times as much energy as all the nuclear weapons on Earth if they were exploded simultaneously.

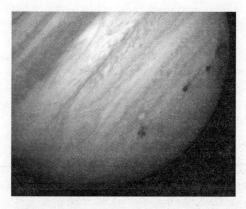

Dark spots are some of the impact sites from
fragments of Shoemaker–Levy 9.

The impacts taught scientists a lot of new things about Jupiter. One is its role as a celestial hoover. Shoemaker–Levy 9 may have been the only comet observed orbiting Jupiter, but at least five others must have done so in the past, judging by their current orbits. All such captures are temporary: either the comet is recaptured by the Sun or it eventually collides with something. Thirteen chains of craters on Callisto and three on Ganymede suggest that sometimes what it hits isn't Jupiter. Put together, the evidence shows that Jupiter sweeps up comets and other cosmic debris, by capturing them and then colliding with them. Such events are rare by our standards, but frequent on a cosmic timescale: a comet 1·6 km across hits Jupiter every 6000 years or so, with smaller ones colliding more often

This aspect of Jupiter helps to protect the inner planets from comet and asteroid impacts, leading to the suggestion in Peter Ward and Donald Brownlee's *Rare Earth*[3] that a large planet like Jupiter makes its inner worlds more habitable for life. Unfortunately for this seductive line of reasoning, Jupiter also disturbs asteroids from the main belt and these can collide with inner planets. If Jupiter were slightly smaller, its overall effect would be detrimental to life on Earth.[4] At its current size, there seems to be no significant overall advantage for Earthly life. *Rare Earth* is ambivalent about impacts in any case: it hails Jupiter as our saviour from comets, while praising its tendency to fling asteroids around as a way to shake up ecosystems and encourage more rapid evolution.

Shoemaker–Levy 9 brought home to many American congressmen the extraordinary violence of a comet impact. The largest impact scar on Jupiter was the same size as the Earth. There's no way we could protect ourselves against an impact of this magnitude with current or foreseeable technology, but it did rather focus the mind on lesser impacts, be they from a comet or an asteroid, where we might be able to prevent a collision if we took steps to give ourselves enough prior warning. Congress rapidly instructed NASA to catalogue all near-Earth asteroids more than a kilometre across. So far 872 have been detected, of which 153 might potentially hit us. Estimates suggest another 70 or so exist, but haven't yet been spotted.

Chaos in the Cosmos

This is highly irregular.

Airplane II: The Sequel

PLUTO'S MOONS ARE WOBBLY.

Pluto has five satellites. Charon is spherical and unusually large compared to its primary, while Nix, Hydra, Kerberos, and Styx are tiny irregular lumps. Charon and Pluto are tidally locked so that each presents the same face to the other. Not so the other moons. In 2015, the Hubble telescope observed irregular variations in the light reflected from Nix and Hydra. Using a mathematical model of spinning bodies, astronomers deduced that these two moons must be tumbling end over end, but not in a nice regular way. Instead, their motion is chaotic.[1]

In mathematics, 'chaotic' is not just a fashionable word for 'erratic and unpredictable'. It refers to *deterministic* chaos, which is apparently irregular behaviour resulting from entirely regular laws. That probably sounds paradoxical, but the combination is often unavoidable. Chaos looks random – and in certain respects it is – but it stems from the same mathematical laws that produce regular, predictable behaviour like the Sun rising every morning.

Further Hubble measurements suggest that Styx and Kerberos also spin chaotically. One of the tasks carried out by *New Horizons* when it visited Pluto was to verify this theory. Its data were to be transmitted back to Earth over a 16-month period, and as I write, the results haven't yet arrived.

Pluto's wobbly moons are the breaking news on chaotic dynamics in the cosmos, but astronomers have discovered many examples of cosmic chaos, from fine details about tiny moons to the long-term future of the solar system. Saturn's moon Hyperion is another chaotic tumbler – the first satellite to be caught behaving badly. The Earth's axis is tilted at a fairly stable 23·4 degrees, giving us the regular succession of seasons, but the axial tilt of Mars varies chaotically. Mercury and Venus used to be like that too, but tidal effects from the Sun have stabilised them.

There's a link between chaos and the 3:1 Kirkwood gap in the asteroid belt. Jupiter clears out asteroids from this region, flinging them willy-nilly around the solar system. Some cross the orbit of Mars, which can redirect them almost anywhere. Maybe that's why the dinosaurs met their demise. Jupiter's Trojan asteroids were probably captured as a consequence of chaotic dynamics. Chaotic dynamics has even provided a way for astronomers to estimate the age of a family of asteroids.

Far from being a gigantic clockwork machine, the solar system plays roulette with its planets. The first hint along these lines, found by Gerry Sussman and Jack Wisdom in 1988, was the discovery that Pluto's orbital elements vary erratically as a consequence of the gravitational forces exerted on it by the other planets. A year later, Wisdom and Laskar showed that the Earth's orbit is also chaotic, though in a milder way: the orbit itself doesn't greatly change, but the Earth's position along the orbit is unpredictable in the long term – 100 million years from now.

Sussman and Wisdom also showed that if there were no inner planets, Jupiter, Saturn, Uranus, and Neptune would behave chaotically in the long term. These outer planets have a significant effect on all the other planets, making them the main source of chaos in the solar system. However, chaos isn't confined to our celestial backyard. Calculations indicate that many exoplanets around far stars probably follow chaotic orbits. There is astrophysical chaos: the light output of some stars varies chaotically.[2] The motion of stars in galaxies may well be chaotic, even though astronomers usually model their orbits as circles (see Chapter 12).

Chaos, it seems, rules the cosmos. Yet astronomers have found that, more often than not, the main cause of chaos is resonant orbits, simple numerical patterns. Like that 3:1 Kirkwood gap. On the other hand, chaos is also responsible for patterns – the spirals of galaxies may well be an example, as we'll also see in Chapter 12.

Order creates chaos, and chaos creates order.

✦

Random systems have no memory. If you roll a dice[3] twice, the number that turns up on the first throw tells you nothing about what will happen on the second throw. It might be the same number as before; then again, it might not be. Don't believe anyone who tries to tell you that if a dice hasn't thrown a 6 for a long time, then the 'law of averages' makes a 6 more likely. There's no such law. It's true that in the long run the proportion of 6s for fair dice should be very close to 1/6, but that happens because large numbers of new tosses swamp any discrepancies, not by the dice suddenly deciding to catch up to where a theoretical average says they ought to be.[4]

Chaotic systems, in contrast, do have a kind of short-term memory. What they're doing now provides hints about what they will do a little into the future. Ironically, if dice were chaotic, then not having thrown a 6 for a long time would be evidence that it probably *won't* happen soon.[5] Chaotic systems have lots of approximate repetitions in their behaviour, so the past is a reasonable – though far from foolproof – guide to the near future.

The length of time for which this kind of forecasting remains valid is called the prediction horizon (jargon: Liapunov time). The more accurately you know the current state of a chaotic dynamical system, the longer the prediction horizon becomes – but the horizon increases far more slowly than the precision of the measurements. However precise they are, the slightest error in the current state eventually grows so large that it overwhelms the prediction. The meteorologist Edward Lorenz discovered this behaviour in a simple model motivated by weather, and the same is true of the sophisticated weather models used by forecasters. The movement of the atmosphere obeys specific mathematical rules with no element of randomness, yet we all know how unreliable weather forecasts can become after just a few days.

This is Lorenz's famous (and widely misunderstood) butterfly effect: a flap of a butterfly's wing can cause a hurricane a month later, halfway round the world.[6]

If you think that sounds implausible, I don't blame you. It's true, but only in a very special sense. The main potential source of misunderstanding is the word 'cause'. It's hard to see how the tiny amount of energy in the flap of a wing can create the huge energy in a hurricane. The answer is, it doesn't. The energy in the hurricane doesn't come from the flap: it's redistributed from elsewhere, when the flap interacts with the rest of the otherwise unchanged weather system.

After the flap, we don't get exactly the same weather as before except for an extra hurricane. Instead, the entire pattern of weather changes, worldwide. At first the change is small, but it grows – not in energy, but in *difference* from what it would otherwise have been. And that difference rapidly becomes large and unpredictable. If the butterfly had flapped its wings two seconds later, it might have 'caused' a tornado in the Philippines instead, compensated for by snowstorms over Siberia. Or a month of settled weather in the Sahara, for that matter.

Mathematicians call this effect 'sensitive dependence on initial conditions'. In a chaotic system, inputs that differ very slightly lead to outputs that differ by large amounts. This effect is real, and very common. For example, it's why kneading dough mixes the ingredients thoroughly. Every time the dough is stretched, nearby grains of flour move further apart. When the dough is then folded over to stop it escaping from the kitchen, grains that are far apart may (or may not) end up close together. Local stretching, combined with folding, creates chaos.

That's not just a metaphor: it's a description in ordinary language of the basic mathematical mechanism that generates chaotic dynamics. Mathematically, the atmosphere is like the dough. The physical laws that govern the weather 'stretch' the state of the atmosphere locally, but the atmosphere doesn't escape from the planet, so its state 'folds back' on itself. Therefore, *if* we could run the Earth's weather twice, with the only difference being an initial *flap* or *no-flap*, the resulting behaviours would diverge exponentially. The weather would still look like weather, but it would be different weather.

In reality we can't run real weather twice, but this is precisely what happens in weather forecasts using models that reflect genuine atmospheric physics. Very tiny changes to the numbers representing the current state of the weather, when input into the equations that predict the future state, lead to large-scale changes in the forecast. For example, an area of high pressure over London in one simulation can be replaced by an area of low pressure in another. The current way round this annoying effect is to run many simulations with small random variations in initial conditions, and use the results to quantify how probable different predictions are. That's what '20% chance of thunderstorms' means.

In practice it's not possible to cause a specific hurricane by employing a suitably trained butterfly, because forecasting the effect of the flap is also subject to the same prediction horizon. Nevertheless, in other contexts, such as the heartbeat, this kind of 'chaotic control' can provide an efficient route to desired dynamic behaviour. We'll see several astronomical examples in Chapter 10, in the context of space missions.

✦

Not convinced? A recent discovery about the early solar system puts the issue into sharp relief. Suppose some celestial superpower could run the formation of the solar system from a primordial gas cloud again, using exactly the same initial state except for *one extra molecule* of gas. How different would today's solar system be?

Not a lot, you might think. But remember the butterfly effect. Mathematicians have proved that bouncing molecules in a gas are chaotic, so it wouldn't be a surprise if the same were true of collapsing gas clouds, even though the details are technically different. To find out, Volker Hoffmann and coworkers simulated the dynamics of a disc of gas at a stage when it contains 2000 planetesimals, keeping track of how collisions cause these bodies to aggregate into planets.[7] They compared the results with simulations including two gas giants, with two distinct choices for their orbits. They made a dozen runs for each of these three scenarios, with slightly different initial conditions. Each run took about a month on a supercomputer.

They found that planetesimal collisions are chaotic, as expected. The butterfly effect is dramatic: change the initial position of a single planetesimal by just one millimetre, and you get a completely different system of planets. Extrapolating from this result, Hoffmann thinks that by adding a single molecule of gas to an exact model of the nascent solar system (were such a thing possible) you'd change the outcome so much that the Earth fails to form.

So much for the clockwork universe.

Before we get carried away by how incredibly unlikely this makes our existence, and invoking the divine hand of providence, we should take into account another aspect of the calculations. Although each run leads to planets with different sizes and different orbits, *all* of the solar systems that arise for a given scenario are very similar to each other. Without any gas giants, we get about 11 rocky worlds, most of them smaller than the Earth. Add the gas giants – a more realistic model – and we get four rocky planets, with masses between half that of the Earth and a bit more than that of the Earth. That's very close to what we actually have. Although the butterfly effect changes orbital elements, the overall structure is much the same as before.

The same happens in weather models. *Flap* ... and the global weather is different from what it would have been – but it's still *weather*. You don't suddenly get floods of liquid nitrogen or a blizzard of giant frogs. So although our solar system would not have arisen in *exactly* its present form if the initial gas cloud had been the slightest bit different, something remarkably similar would have arisen instead. So living organisms would probably have been just as likely to evolve.

The prediction horizon can sometimes be used to estimate the age of a chaotic system of celestial bodies, because it governs how fast the system breaks up and disperses. Asteroid families are examples. They can be spotted because their members have very similar orbital elements. Each family is thought to have been created by the break-up of a single larger body at some time in the past. In 1994 Andrea Milani and Paolo Farinella used this method to deduce that the Veritas asteroid family is at most 50 million years old.[8] This is a compact cluster of asteroids associated with 490 Veritas, towards the outside of the main belt and just inside the 2:1 resonant orbit with Jupiter. Their calculations show that two of the asteroids in this family have strongly chaotic orbits, created by a temporary 21:10 resonance with Jupiter. The prediction horizon implies that these two asteroids should not have stayed close together for more than 50 million years, and other evidence suggests they are both original members of the Veritas family.

✦

The first person to recognise the existence of deterministic chaos, and to gain some inkling of why it happens, was the great mathematician Henri Poincaré. He was competing for a mathematical prize offered by King Oscar II of Norway and Sweden, asking for a solution of the n-body problem for Newtonian gravitation. The rules for the prize specified what sort of solution was required. Not a formula like Kepler's ellipse, because everyone was convinced no such thing existed, but 'a representation of the coordinates of each point as [an infinite] series in a variable that is some known function of time and for all of whose values the series converges uniformly'.

Poincaré discovered that the task is essentially impossible, even for three bodies under very restrictive conditions. The way he proved it was to demonstrate that orbits can be what we now call 'chaotic'.

The general problem for any number of bodies proved too much even for Poincaré. He took $n = 3$. In fact, he worked on what I called the $2\frac{1}{2}$-body problem in Chapter 5. The two bodies are, say, a planet and one of its moons; the half-body is a grain of dust, so lightweight that although it responds to the gravitational fields of the other two bodies, it has absolutely no effect on *them*. What emerges from this model is a lovely combination of perfectly regular two-body dynamics for the massive bodies and highly erratic behaviour for the dust particle. Ironically, it's the regular behaviour of the massive bodies that makes the dust particle go crazy.

'Chaos' makes it sound as though the orbits of three or more bodies are random, structureless, unpredictable, and lawless. Actually, the dust particle loops round and round in smooth paths close to arcs of ellipses, but the shape of the ellipse keeps changing without any obvious pattern. Poincaré came across the possibility of chaos when he was thinking about the dynamics of the dust grain when it happens to be close to a periodic orbit. What he expected was some complicated combination of periodic motions with different periods, much as an orbiting capsule goes round the Moon goes round the Earth goes round the Sun – all in different periods of time. However, as already specified in the rules for the prize, the answer was expected to be a 'series', which combines infinitely many periodic motions, not just three.

Poincaré found such a series. How, then, does chaos appear? Not as a consequence of the series, but because of a flaw in the whole idea. The rules stated that the series must *converge*. This is a technical mathematical requirement for an infinite sum to make sense. Essentially, the sum of the series should get closer and closer to some specific number as you include more and more terms. Poincaré was alert to pitfalls, and realised that his series didn't converge. At first, it seemed to be getting closer and closer to some specific number, but then the sum started to diverge from that number by ever greater amounts. This behaviour is characteristic of an 'asymptotic' series. Sometimes an asymptotic series is useful for practical purposes, but here it hinted at an obstacle to obtaining a genuine solution.

To figure out what that obstacle was, Poincaré abandoned formulas and series, and turned to geometry. He considered both position and velocity, so the contour lines in the picture on page 79 are really three-dimensional objects, not curves. This causes extra complications. When he thought about the geometric arrangement of all the possible orbits near a particular periodic one, he realised that many orbits must be very tangled and erratic. The reason lay in a special pair of curves, which captured how nearby orbits either approach the periodic one or diverge from it. If these curves cross each other at some point, then basic mathematical features of dynamics (uniqueness of solutions of a differential equation for given initial conditions) imply that they must cross at infinitely many points, forming a tangled web. Soon after, in *Les Méthodes Nouvelles de la Mécanique Celeste* (New Methods of Celestial Mechanics) he described the geometry as:

> a kind of trellis, a fabric, a network of infinitely tight mesh; each of the two curves must not cross itself but it must fold on itself in a very complicated way to intersect all of the meshes of the fabric infinitely many times. One will be struck by the complexity of this picture, which I will not even attempt to draw.

Today we call this picture a homoclinic tangle. Ignore 'homoclinic' (jargon: an orbit that joins an equilibrium point to itself) and focus on 'tangle', which is more evocative. The picture on page 127 explains the geometry in a simple analogue.

Ironically, Poincaré very nearly missed making this epic discovery. While looking through documents in the Mittag-Leffler Institute in Oslo, the mathematical historian June Barrow-Green discovered that the published version of his prizewinning work was not the one he'd submitted.[9] After the prize had been awarded and the official memoir had been printed but not yet distributed, Poincaré had discovered a mistake – he'd overlooked chaotic orbits. He withdrew his memoir and paid for a revised 'official' version to be quietly substituted.

✦

It took a while for Poincaré's new ideas to sink in. The next big advance came in 1913 when George Birkhoff proved the 'Last Geometric Theorem', an unproved conjecture that Poincaré had used to deduce the occurrence, in suitable circumstances, of periodic orbits. We now call this result the Poincaré–Birkhoff fixed point theorem.

Mathematicians and other scientists became fully aware of chaos about fifty years ago. Following in Birkhoff's footsteps, Stephen Smale made a deeper study of the geometry of the homoclinic tangle, having encountered the same problem in another area of dynamics. He invented a dynamical system with much the same geometry that's easier to analyse, known as the Smale horseshoe. This system starts with a square, stretches it out into a long thin rectangle, folds it round in a horseshoe shape, and fits it back on top of the original square. Repeating this transformation is much like kneading dough, and it has the same chaotic consequences. The horseshoe geometry allows a rigorous proof that this system is chaotic, and that in some respects it behaves like a random sequence of coin tosses – despite being completely deterministic.

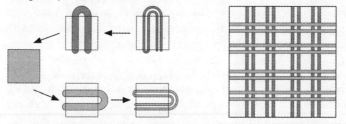

Left: Smale's horseshoe. The square is repeatedly folded, creating a series of horizontal stripes. Reversing time and unfolding it converts these into similar vertical stripes. Right: When the two sets of stripes cross, we get a homoclinic tangle. The dynamics – obtained by repeatedly folding – makes points jump around on the tangle, apparently at random. The complete tangle involves infinitely many lines.

As the extent and richness of chaotic dynamics became apparent, the growing excitement triggered a lot of interest from the media, who dubbed the whole enterprise 'chaos theory'. Really, the topic is one part – a significant and fascinating part, to be sure – of an even more important area of mathematics, known as nonlinear dynamics.

✦

The strange behaviour of Pluto's moons is just one example of chaos in the cosmos. In 2015 Mark Showalter and Douglas Hamilton published a mathematical analysis backing up the Hubble's puzzling observations of the moons of Pluto.[10] The idea is that Pluto and Charon act like the dominant bodies in Poincaré's analysis, and the other, far smaller, moons act a bit like the dust particle. However, because they're not point particles, but shaped like rugby footballs, or possibly even potatoes, their craziness shows up as chaotic tumbling. Their orbits, and where the moons will be in those orbits at any given time, are also chaotic: predictable only statistically. Even less predictable is the direction in which each moon will point.

Pluto's moons weren't the first tumblers to be spotted. That honour goes to Saturn's satellite Hyperion, and at the time it was thought to be the only tumbling moon. In 1984 Hyperion attracted the attention of Wisdom, Stanton Peale, and François Mignard.[11] Almost all moons in the solar system fall into two categories. The axial spin of a moon in the first category has been heavily modified by tidal interactions with its parent planet, so it always presents the same face to the planet, a 1:1 spin–orbital resonance, otherwise known as synchronous rotation. For the second category, very little interaction has taken place and it still spins much as it did when it first formed. Hyperion and Iapetus are exceptions: according to theory, they should eventually lose most of their initial spin and synchronise it with their orbital revolution, but not for a long time – about a billion years.

Despite that, Iapetus already rotates synchronously. Hyperion alone seemed to be doing something more interesting. The question was: what?

Wisdom and his colleagues compared data on Hyperion to a theoretical criterion for chaos, the resonance overlap condition. This predicted that Hyperion's orbit should interact chaotically with its spin, a prediction confirmed by solving the equations of motion numerically. The chaos in Hyperion's dynamics manifests itself mainly as erratic tumbling. The orbit itself doesn't vary as wildly. It's like an American football rolling round and round an athletics track, sticking to one lane but tumbling unpredictably end over end.

In 1984 the only known moon of Pluto was Charon, discovered in 1978, and no one could measure its rate of spin. The other four were discovered between 2005 and 2012. All five are crammed into an unusually small zone, and it's thought that they were originally all part of a single larger body, which collided with Pluto during the early formation of the solar system – a miniature version of the giant impact theory of the formation of our own Moon. Charon is large, round, and tidally locked in a 1:1 resonance, so it always presents the same face to Pluto, just as the Moon does to the Earth. However, unlike the Earth, Pluto also always presents the same face to its moon. The tidal locking, and the round shape, prevent chaotic tumbling. The other four moons are small, irregular, and are now known to tumble chaotically, like Hyperion.

Plutonian numerology doesn't stop with that 1:1 resonance. To a good approximation, Styx, Nix, Kerberos, and Hydra are in 1:3, 1:4, 1:5, and 1:6 orbital resonances with Charon; that is, their periods are roughly 3, 4, 5, and 6 times as long as Charon's. However, those figures are only averages. The actual orbital periods vary significantly from one revolution to the next.

Even so, in astronomical terms it all looks very orderly. Because order can give rise to chaos, it's common for them both to coincide in the same system: orderly in some respects, chaotic in others.

✦

The two main research groups that work on chaos and the long-term dynamics of the solar system are headed by Wisdom and Laskar. In 1993, within a week of each other, both groups published papers describing a new cosmic context for chaos: the axial tilt of the planets.

In Chapter 1 we saw that a rigid body spins about an axis: a line through the body that is instantaneously stationary. The spin axis can move over time, but in the short term it stays pretty much fixed. So the body spins like a top, with the axis as the central spindle. Planets, being almost spherical, spin at a very regular rate about an axis that seems not to change, even over centuries. In particular, the angle between the axis and the ecliptic plane, technically known as the obliquity, remains constant. For the Earth it's 23·4 degrees.

However, appearances are deceptive. Around 160 BC Hipparchus discovered an effect known as the precession of the equinoxes. In the *Almagest*, Ptolemy states that Hipparchus observed the positions in the night sky of the star Spica (alpha Virginis) and others. Two predecessors had done the same: Aristillus around 280 BC and Timocharis around 300 BC. Comparing data, Ptolemy concluded that Spica had drifted by about two degrees when observed at the autumnal equinox – the time at which night and day are equally long. He deduced that the equinoxes were moving along the zodiac at about one degree every century, and would eventually get back to where they started after 36,000 years.

We now know he was right, and why. Rotating bodies precess: their spin axis slowly changes direction, as the tip of the axis describes a slow circle. Spinning tops often do this. Mathematics going back to Lagrange explains precession as the typical dynamics of a body with a certain type of symmetry – two equal axes of inertia. Planets are approximately ellipsoids of rotation, so they satisfy this condition. The Earth's axis precesses with a period of 25,772 years. This affects how we see the night sky. At the moment, the pole star, Polaris, in Ursa Major is aligned with the axis and it therefore appears to be fixed, while the rest of the stars seem to rotate around it. Actually, it's the Earth that's rotating. But in ancient Egypt, 5000 years ago, Polaris went round in a circle, and the faint star Batn al Thuban (phi Draconis) was fixed instead. I chose that date because it's a matter of luck whether there's a bright star near the pole or not, and mostly there isn't.

When a planet's axis precesses, its obliquity doesn't change. The seasons slowly drift, but so slowly that only a Hipparchus would notice, and only then with the help of previous generations. A given location on the planet experiences much the same seasonal variations, but their timing changes very slowly. Both Laskar's and Wisdom's groups discovered that Mars is different. Its obliquity also varies, to some extent driven by changes in its orbit. If the precession of its axis resonates with the period of any variable orbital element, the obliquity can change. The two groups calculated what effect this has by analysing the planet's dynamics.

Wisdom's calculations show that the obliquity of Mars varies chaotically, ranging between 11 and 49 degrees. It can change by 20 degrees in about 100,000 years, and it oscillates chaotically over that sort of range at about that rate. Nine million years ago the obliquity varied between 30 and 47 degrees, and this continued until 4 million years ago, when there was a relatively abrupt shift to a range between 15 and 35 degrees. The calculations include effects from general relativity, which in this particular problem are important. Without those, the model doesn't lead to this transition. The reason for the transition is – you've guessed it – passage through a spin–orbit resonance.

Laskar's group used a different model, without relativistic effects but with a more accurate representation of the dynamics, and examined a longer period of time. The group obtained similar results for Mars, but found that over longer periods its obliquity varies between 0 and 60 degrees, an even wider range.

They also studied Mercury, Venus, and Earth. Today, Mercury spins very slowly, once every 58 days, and it goes round the Sun in 88 days – a 3:2 spin–orbit resonance. This was probably caused by tidal interactions with the Sun, which slowed the primordial spin down. Laskar's group calculated that orginally Mercury spun once every 19 hours. Before the planet reached its current state, its obliquity varied between 0 and 100 degrees, taking about a million years to cover most of that range. In particular, there were times when its pole faced the Sun.

Venus poses a puzzle for astronomers, because, by the usual conventions about angles for spinning bodies, its obliquity is 177 degrees – essentially upside down. This causes it to rotate very slowly (period 243 *days*) in the opposite direction to every other planet. The explanation for this 'retrograde' motion isn't known, but in the 1980s it was thought to be primordial: going right back to the origin of the solar system. Laskar's analysis suggests this might not be the case. It's thought that Venus originally had a rotation period of a mere 13 hours. Assuming this, the model shows that the obliquity of Venus originally varied chaotically, and when it reached 90 degrees it could have become stable rather than chaotic. From that state, it could gradually evolve to its present value.

The results for the Earth are interestingly different. Earth's obliquity is very stable, varying by only a degree. The reason seems to be our unusually large Moon. Without it, Earth's obliquity would wander around between 0 and 85 degrees. On this alternative Earth, climatic conditions would be very different. Instead of the equator being warm and the poles cold, different regions would experience entirely different ranges of temperature. This would affect the weather patterns.

Some scientists have suggested that without the Moon the chaotic changes in climate would have made it harder for life, especially complex life, to evolve here. However, life evolved in the oceans. It didn't invade the land until about 500 million years ago. Marine life would not be greatly affected by a changing climate. As for land animals, the climatic changes that would result from the absence of the Moon are fast on astronomical timescales, but land organisms would migrate as the climate changed, because on their timescale the changes are slow. Evolution would proceed largely unhindered. It might even be speeded up by stronger pressure to adapt.

✦

Astronomical effects on Earth's living creatures that actually happened are more interesting than hypothetical ones that didn't. The most famous is the asteroid that destroyed the dinosaurs. Or was it a comet? And were other influences involved too, such as massive volcanic eruptions?

Dinosaurs first appeared about 231 million years ago in the Triassic, and disappeared 65 million years ago at the end of the Cretaceous. In between, they were the most successful vertebrates, in sea and on land. By comparison, 'modern' humans have been around for about 2 million years. However, there were many species of dinosaur, so that's a bit unfair. Most individual species survive for no more than a few million years.

The fossil record shows that the dinosaurs died out very suddenly by geological standards. Their demise was accompanied by that of mosasaurs, plesiosaurs, ammonites, many birds, most marsupials, half the types of plankton, many fishes, sea urchins, sponges, and snails. This 'K/T extinction' is one of five or six major events in which huge numbers of species perished in a geological eye-blink.[12] The dinosaurs did manage to leave some modern descendants, though: birds evolved from theropod dinosaurs in the Jurassic. Towards the end of their reign dinosaurs coexisted with mammals, some quite large, and the disappearance of dinosaurs seems to have triggered a burst of mammalian evolution as the main competition was removed from the scene.

There's a widespread consensus among palaeontologists that a major cause of the K/T extinction was the impact of an asteroid, or possibly a comet, which left an indelible mark on the Yucatan coast of Mexico: the Chicxulub crater. Whether this was the sole cause is still contentious, partly because there's at least one other plausible candidate: the massive volcanic outpourings of magma that formed the Deccan Traps in India, which would have sent large amounts of noxious gases into the atmosphere. 'Traps' here comes from the Swedish for 'stairs' – the basalt strata tend to weather into a series of steps. Maybe climate change or changing sea levels were involved too. But the impact is still the prime suspect, and several attempts to prove otherwise have foundered as improved evidence came in.

The main problem with the Deccan Traps theory, for instance, is that they formed over a period of 800,000 years. The K/T extinction was much more rapid. In 2013 Paul Renne used argon–argon dating (a comparison of proportions of different isotopes of the gas argon) to pin down the impact to 66·043 million years ago, plus or minus 11,000 years. The death of the dinosaurs seems to have happened within 33,000 years of that date. If correct, the timing seems too close to be a coincidence. But it's certainly possible that other causes stressed the world's ecosystem, and the impact was the *coup de grâce*. In fact, in 2015 a team of geophysicists led by Mark Richards found clear evidence that shortly after the impact, the flow of lava from the Deccan traps doubled.[13] This adds weight to an older theory: the impact sent shockwaves round the Earth. They focused on the region diametrically opposite Chicxulub, which happens to be very close to the Deccan traps.

Astronomers have tried to find out whether the impactor was a comet or an asteroid, and even where it came from. In 2007 William Bottke and others[14] published an analysis of chemical similarities suggesting that the impactor originated in a group of asteroids known as the Baptinista family, and that this broke up about 160 million years ago. But at least one asteroid from this group has the wrong chemistry, and in 2011 the timing of the break-up was estimated as 80 million years, which doesn't leave a long enough gap before the impact.

✦

One thing that has been established is how chaos causes asteroids to be flung out of their belt and end up hitting the Earth. The culprit is Jupiter, ably assisted by Mars.

Recall from Chapter 5 that the asteroid belt has gaps – distances from the Sun where the population is unusually sparse – and that these correlate well with orbits in resonance with Jupiter. In 1983 Wisdom studied the formation of the 3:1 Kirkwood gap, seeking to understand the mathematical mechanism that causes asteroids to be ejected from such an orbit. Mathematicians and physicists had already discovered a close association between resonance and chaos. At the heart of a resonance is a periodic orbit, in which the asteroid makes a whole number of revolutions while Jupiter makes another whole number. Those two numbers characterise the resonance, and in the above example they are 3 and 1. However, such an orbit will change because other bodies perturb the asteroid. The question is: how?

In the mid twentieth century three mathematicians – Andrei Kolmogorov, Vladimir Arnold, and Jürgen Moser – obtained different bits of the answer to this question, collected together in the KAM theorem. This states that orbits near the periodic one are of two kinds. Some are quasiperiodic, spiralling around the original orbit in a regular manner. The others are chaotic. Moreover, the two types are nested in an intricate manner. The quasiperiodic orbits spiral around tubes that surround the periodic orbit. There are infinitely many of these tubes. Between them are more complicated tubes, spiralling around the spiral orbits. Between those are even more complicated tubes spiralling around *those*, and so on. (This is what 'quasiperiodic' means.) The chaotic orbits fill the intricate gaps between all of these spirals and multiple spirals, and are defined by Poincaré's homoclinic tangles.

This highly complex structure can most easily be visualised by borrowing a trick from Poincaré and looking at them in cross section. The initial periodic orbit corresponds to the central point, the quasiperiodic tubes have the closed curves as cross sections, and the shaded regions between them are traces of chaotic orbits. Such an orbit passes through some point in the shaded region, travels all the way round near the original periodic orbit, and hits the cross section again at a second point – whose relation to the first appears random. What you'd observe wouldn't be an asteroid performing a drunkard's walk; it would be an asteroid whose orbital elements change chaotically from one orbit to the next.

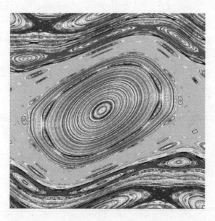

Numerically computed cross section of orbits near a periodic one, in accordance with the KAM theorem.

To carry out specific computations for the 3:1 Kirkwood gap, Wisdom invented a new method to model the dynamics: a formula that matches how successive orbits hit the cross section. Instead of solving a differential equation for the orbit, you just keep applying the formula. The results confirm that chaotic orbits occur, and provide details of what they look like. For the most interesting ones, the eccentricity of the approximate ellipse suddenly grows much larger. So an orbit that's reasonably close to a circle, maybe a fattish ellipse, turns into a long thin one. Long enough, in fact, to cross the orbit of Mars. Since it keeps doing that, there's good chance that it will come close to Mars, and be perturbed by the slingshot effect. And that would fling it … anywhere. Wisdom suggested that this mechanism is how Jupiter clears out the 3:1 Kirkwood gap. As confirmation, he plotted the orbital elements of asteroids near the gap, and compared them to the chaotic zone of his model. The fit is near perfect.

Basically, an asteroid trying to orbit in the gap gets shaken up by the chaos and gets passed to Mars, which kicks it away. Jupiter takes a

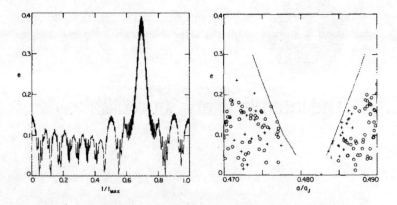

Left: Spike in eccentricity (vertical axis). Horizontal axis is time. Right: Outer edges of the chaotic zone (solid lines) and orbital elements of asteroids (dots and crosses). Vertical axis is eccentricity, horizontal axis is major radius relative to that of Jupiter.

corner kick, Mars scores. And sometimes ... just sometimes ... Mars kicks it in our direction. And if the kick happens to be on target –

Mars one, dinosaurs nil.

The Interplanetary Superhighway

Space travel is utter bilge.

Richard Woolley, Astronomer Royal, 1956

WHEN VISIONARY SCIENTISTS and engineers first started thinking seriously about landing humans on the Moon, one of the first problems was to work out the best route. 'Best' has many meanings. In this instance the requirements are a fast trajectory, minimising the time vulnerable astronauts spend hurtling through vacuum in a glorified tin can, and switching the rocket engine on and off as few times as possible to reduce the chance of it failing.

When *Apollo 11* landed two astronauts on the Moon, its trajectory obeyed these two principles. First, the spacecraft was injected into low Earth orbit, where everything could be checked to make sure it was still functional. Then a single burst of the engines sent it speeding towards the Moon. When it got close, a few more bursts slowed it down again, injecting it into lunar orbit. The landing module then went down to the surface, and its top half came back a few days later with the crew. It was then jettisoned, and the crew returned to Earth with another burst from the engine to take them out of lunar orbit. After coasting home they came to the most dangerous part of the entire mission: using friction with the Earth's atmosphere as a brake, to slow the command capsule down enough for it to land using parachutes.

For a time, this type of trajectory, which in its simplest form is known as a Hohmann ellipse, was used for most missions. There's a sense in which the Hohmann ellipse is optimal. Namely, it's faster than most alternatives, for the same amount of rocket fuel. But as humanity

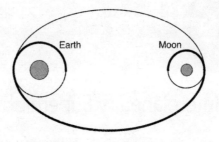

Hohmann ellipse. Thick line shows transfer orbit.

gained experience with space missions, engineers realised that other types of mission have different requirements. In particular, speed is less important if you're sending a machine or supplies.

Until 1961 mission planners, convinced that a Hohmann ellipse is optimal, viewed the gravitational field of a planet as an obstacle, to be overcome using extra thrust. Then Michael Minovitch discovered the slingshot effect in a simulation.[1] Within a few decades, new ideas from the mathematics of many-body orbits led to the discovery that a spacecraft can get to its destination using far less fuel, by following a trajectory very different from that used for the Moon landing. The price is that it takes much longer, and may require a more complex series of rocket boosts. However, today's rocket engines are more reliable, and can be fired repeatedly without greatly increasing the likelihood of failure.

Instead of considering just the Earth and the final target, engineers started thinking about all of the bodies that might potentially affect a space probe's trajectory. Their gravitational fields combine to create a kind of energy landscape, a metaphor that we encountered in connection with Lagrange points and Greek and Trojan asteroids. The spacecraft in effect wanders around the contours of this landscape. One twist is that the landscape changes as the bodies move. Another is that mathematically this is a landscape in many dimensions, not just the usual three, because velocity is important as well as position. A third is that chaos plays a key role: you can take advantage of the butterfly effect to obtain large results from small causes.

These ideas have been used in real missions. They also imply that

Artist's conception of the interplanetary superhighway. The
ribbon represents one possible trajectory along a tube,
and constrictions represent Lagrange points.

the solar system has a network of invisible mathematical tubes linking
its planets, an interplanetary superhighway system providing unusually
efficient routes between them.[2] The dynamics governing these tubes
may even explain how the planets are spaced, a modern advance on the
Titius–Bode law.

✦

The *Rosetta* mission is an example of new ways to design trajectories for space probes. It doesn't use the butterfly effect, but it shows
how imaginative planning can produce results that at first seem impossible, by exploiting natural features of the solar system's gravitational
landscape. *Rosetta* was technically challenging, not least because of
the distance and speed of the target. At the time of the landing, comet
67P was 480 million kilometres from Earth and travelling at over 50,000

kph. That's sixty times as fast as a passenger jet aircraft. Because of the limitations of current rocketry, the point-and-go method used for the Moon landing won't work.

Getting out of Earth orbit with enough speed is difficult and expensive, but it's possible. Indeed, the *New Horizons* mission to Pluto took the direct route. It did borrow some extra velocity from Jupiter along the way, but it could have got there without that by taking longer. The big problem was slowing down again; this was solved by not even trying. *New Horizons*, the fastest space vehicle ever launched, used a very powerful rocket with five solid fuel boosters and an extra final stage to get up to speed when leaving the Earth. It also left them behind as soon as it could: too heavy to keep, and empty of fuel anyway. When the probe got to Pluto it barrelled through the system at high speed, and had to do all of its main scientific observations within a period of about a day. During that time it was too busy to communicate with Earth, causing a nervous period while the mission scientists and controllers waited to see whether it had survived the encounter – colliding with a single dust grain could have proved fatal.

In contrast, *Rosetta* had to rendezvous with 67P and *stay with it* as the comet neared the Sun, observing it all the time. It had to deposit *Philae* on the comet's surface. Relative to the comet, *Rosetta* had to be pretty much stationary, but the comet was 300 million miles away and moving at a colossal speed – 55,000 kph. So the mission trajectory had to be designed to bring it up to speed, yet end up in the same orbit as the comet. Even finding a suitable trajectory was difficult; so was finding a suitable comet.

In the event, the probe followed a highly indirect route,[3] which among other things returned near the Earth *three times*. It was a bit like travelling from London to New York by first shuttling back and forth several times between London and Moscow. But cities stay still relative to the Earth, whereas planets don't, and that makes all the difference. The probe began its epic journey by moving in what naively appears to be totally the wrong direction. It headed *towards* the Sun, even though the comet was far outside the orbit of Mars, and moving away. (I don't mean *directly* towards: just that the distance to the Sun was becoming shorter.) *Rosetta*'s orbit swung past the Sun and returned close to the Earth, where it was flung outwards to an encounter with Mars. It then swung back to meet the Earth for a second time, then back out beyond

Mars's orbit *again*. By now the comet was on the far side of the Sun and closer to it than *Rosetta* was. A third encounter with Earth flung the probe outwards again, chasing the comet as it now sped away from the Sun. Finally, *Rosetta* made its rendezvous with destiny.

Why such a complicated route? ESA didn't just point its rocket at the comet and blast off. That would have required far too much fuel, and by the time it got there, the comet would have been somewhere else. Instead, *Rosetta* performed a carefully choreographed cosmic dance, tugged by the combined gravitational forces of the Sun, the Earth, Mars, and other relevant bodies. Its route, calculated by exploiting Newton's law of gravity, was designed for fuel efficiency. Each close fly-by with Earth and Mars gave the probe a free boost as it borrowed energy from the planet. An occasional small burst from four thrusters kept the craft on track. The price paid for conserving fuel was that *Rosetta* took ten years to get to its destination. However, without paying that price, the mission would have been far too costly to get off the ground at all.

This kind of trajectory, going round and round and in and out, seeking judicious speed boosts from planets and moons, has become commonplace for space missions when time is not of the essence. If a space probe passes close behind a planet as it travels along its orbit, the probe can steal some of the planet's energy in a slingshot manoeuvre. The planet actually *slows down*, but the decrease is too small for even the most sensitive apparatus to observe. So the probe gets a boost in speed without having to use up any rocket fuel.

The devil, as always, is in the detail. In order to design such trajectories, the engineers must be able to predict the movements of all bodies involved, and they have to make the whole journey fit together to get the probe to its intended destination. So mission design is a mixture of calculation and black art. Everything depends on an area of human activity whose role in space exploration is seldom even hinted at, but without which, nothing could be achieved. Whenever the media start talking about 'computer models' or 'algorithms', you can presume that they really mean 'mathematics', but are either too scared to mention the word, or think it will scare *you*. There are sensible reasons not to rub people's noses in complex mathematical detail, but it does a grave disservice to one of humanity's most powerful ways of thinking to pretend it's not there at all.

✦

Rosetta's main dynamic trick was the slingshot manoeuvre. Aside from those repeated encounters, it effectively followed a series of Hohmann ellipses. Instead of going into orbit around 67P, it followed a nearby ellipse round the Sun. But there's a far more intriguing trick, a real game changer, that's revolutionising the design of mission trajectories. Astonishingly, it's based on chaos.

As I explained in Chapter 9, chaos in the mathematical sense is not simply a fancy term for random or erratic behaviour. It's behaviour that *appears* random and erratic, but is actually governed by a hidden system of explicit deterministic rules. For celestial bodies, those rules are the laws of motion and gravity. At first sight, the rules don't greatly help, because their main implication is that chaotic motion is unpredictable in the long term. There's a prediction horizon, beyond which any predicted motion will be swamped by unavoidable tiny errors in measuring the current state. Over the horizon, all bets are off. So chaos looks like a bad thing all round.

One of the early criticisms of 'chaos theory' was that because chaos is unpredictable, it causes difficulties for humans trying to understand nature. What's the point of a theory that makes everything harder? It's worse than useless. Somehow the people making this point seemed to imagine that nature would therefore miraculously arrange itself to avoid chaos and help us out. Or that if we hadn't *noticed* some systems are unpredictable, they would have been predictable instead.

The world doesn't work like that. It feels no compulsion to oblige humans. The job of scientific theories is to help us understand nature; improved control of nature is a common by-product, but it's not the main aim. We know, for example, that the Earth's core consists of molten iron, so there's no serious prospect of going there, even with an autonomous tunnelling machine. What a silly theory! How pointless. Except – sorry, it's true. And actually, it's useful too: it helps to explain the Earth's magnetic field, something that helps keep us alive by deflecting radiation.

Similarly, the main point of chaos theory is that chaos is *there*, in the natural world. In suitable, common, circumstances it's just as inevitable a consequence of the laws of nature as those nice simple patterns like periodic elliptical orbits that kick-started the scientific revolution.

Since it's there, we have to get used to it. Even if the only thing we could do with chaos theory was to warn people to expect erratic behaviour in rule-based systems, that would be worth knowing. It would stop us looking for non-existent external influences that we might otherwise presume to be the cause of the irregularities.

Actually, 'chaos theory' has more useful consequences. Because chaos emerges from rules, you can use the chaos to infer the rules, test the rules, and make deductions from the rules. Since nature often behaves chaotically, we'd better try to understand how chaos operates. But the truth is more positive still. Chaos can be good for you, thanks to the butterfly effect. Small initial differences can cause big changes. Let's turn that around. Suppose you want to cause a hurricane. Sounds like a huge task. But, as Terry Pratchett pointed out in *Interesting Times*, all you need to do is find the right butterfly, and ... *flap*.

This is chaos, not as an obstacle, but as a very efficient form of control. If we could somehow reverse-engineer the butterfly effect, we would be able to redirect a chaotic system to a new state with very little effort. We could topple a government and start a war just by moving a finger. Unlikely? Yes, but remember Sarajevo. If the circumstances are right, one finger on the trigger of a pistol is all it takes.[4]

The many-body problem in astronomy is chaotic. Harnessing the butterfly effect in that context lets us redirect a space probe using hardly any propellant. We might, for example, kick an almost worn-out lunar probe out of its death orbit round the Moon, and send it off to look at a comet. That also sounds unlikely, but in principle the butterfly effect ought to be able to hack it.

What's the snag? (There always is one. No such thing as a free lunch.)

Finding the right butterfly.

✦

A Hohmann ellipse links an Earth orbit to an orbit around the target world, and with a bit of tweaking it's a pretty good choice for manned missions. If you're transporting perishable goods (humans) then you need to get to the destination fast. If time is not of the essence, however, there are alternative routes, which take longer but use less fuel. To exploit the butterfly effect, we need a source of chaos. A Hohmann

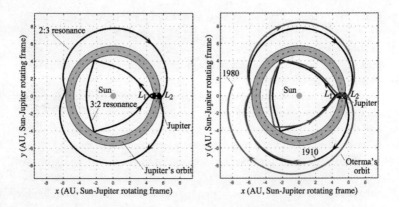

Left: Gravitational landscape for Oterma's orbit, showing periodic orbit in 3:2 resonance with Jupiter. Right: the comet's actual orbit from 1910 to 1980.

ellipse consists of three different two-body orbits (ellipse and circles) patched together, using boosts from propellant to switch the probe from one to another. But there's no chaos in the two-body problem. Where do we find orbital chaos? In the three-body problem. So what we should be thinking about is patching together *three-body* orbits. We can throw in two-body ones as well, if they help, but we're not confined to those.

In the late 1960s Charles Conley and Richard McGehee pointed out that each such path is surrounded by a nested set of tubes, one inside the other. Each tube corresponds to a particular choice of speed; the further from the optimal speed, the wider the tube. On the surface of any given tube, the total energy is constant. It's a simple idea, with a remarkable consequence. To visit another world in a fuel-efficient manner, go by tube.

The planets, moons, asteroids, and comets are tied together by a network of tubes. The tubes have always been there, but they can be seen only through mathematical eyes, and their walls are energy levels. If we could visualise the ever-changing landscape of gravitational fields that controls how the planets move, we'd be able to see the tubes, swirling along with the planets in their slow, stately gravitational dance. But we now know that the dance can be unpredictable.

Take Oterma, for example: a remarkably unruly comet. A century

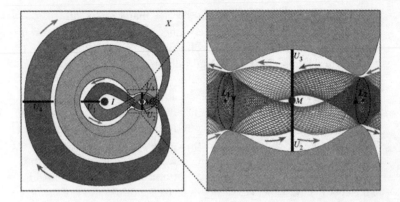

Left: Tube system for Oterma. Right: Close-up of switching region.

ago, Oterma's orbit was well outside the orbit of Jupiter. After a close encounter, its orbit shifted inside Jupiter's. Then it switched back again. Oterma will continue to switch orbits every few decades: not because it breaks Newton's law, but because it obeys it. Oterma's orbit lies inside two tubes, which meet near Jupiter. One tube lies inside Jupiter's orbit, the other outside. At the junction, the comet switches tubes, or not, depending on chaotic effects of Jovian and solar gravity. But once inside a tube, Oterma is stuck there until it returns to the junction. Like a train that has to stay on the rails, but can change its route to another set of rails if someone switches the points, Oterma has some freedom to change its itinerary, but not much.

✦

Victorian railway builders understood the need to exploit natural features of the landscape. They ran railways through valleys and along contour lines; they dug tunnels through hills to avoid running the train over the top. Climbing a hill, against the force of gravity, costs energy. This cost shows up as increased fuel consumption, which in turn costs money. It's the same with interplanetary travel, but the energy landscape changes as the planets move. It has many more dimensions than the two that characterise the location of a train. They represent two distinct physical quantities: location and velocity. A spacecraft travels through a mathematical landscape that has six dimensions rather than two. The tubes and their junctions are special features of the gravitational landscape of the solar system.

A natural landscape has hills and valleys. It takes energy to climb a hill, but a train can gain energy by rolling down into a valley. Two types of energy come into play here. Height above sea level determines the train's potential energy, which traces back to the force of gravity. Then there's kinetic energy, which corresponds to speed. When a train rolls downhill and accelerates, it trades potential energy for kinetic. When it climbs a hill and slows down, the trade is in the reverse direction. The total energy is constant, so the motion runs along a contour line in the energy landscape. However, trains have a third source of energy: fuel. By burning diesel or using electrical power the train can climb a gradient or speed up, freeing itself from its natural free-running trajectory. At any instant, the total energy must remain the same, but all else is negotiable.

It's much the same with spacecraft. The gravitational fields of the Sun, planets, and other bodies provide potential energy. The speed of the spacecraft corresponds to kinetic energy. And its motive power adds a further energy source, which can be switched on or off at will. Energy plays the role of height in the landscape, and the path followed by the spacecraft is a kind of contour line, along which the total energy remains constant. Crucially, you don't have to stick to a single contour: you can burn some fuel to switch to a different one, moving 'uphill' or 'downhill'.

The trick is to do this in the right place. Victorian railway engineers were well aware that the terrestrial landscape has special features – the peaks of hills, the bottoms of valleys, the saddle-shaped geometry of mountain passes. These features are important: they form a kind of skeleton for the overall geometry of the contours. For instance, near a peak the contours form closed curves. At peaks, potential energy is locally at a maximum, in a valley it's at a local minimum. Passes combine features of both – maximum in some directions, minimum in others – and they get you past the mountains with the least expenditure of effort.

Similarly, the energy landscape of the solar system has special features. The most obvious are the Sun, planets, and moons, which sit at the bottom of gravity wells. Equally important, but less visible, are the hilltops, valley bottoms, and mountain passes of the energy landscape. These features organise the overall geometry, and it's this geometry that creates the tubes. The best-known features of the energy landscape, other than gravity wells, are the Lagrange points.

Around 1985 Edward Belbruno pioneered the use of chaotic dynamics in mission planning, introducing what was then called fuzzy boundary theory. He realised that when coupled to chaos, the tubes determine new energy-efficient routes from one world to another. The routes are built from pieces of natural orbits in three-body systems, which have new features like Lagrange points. One way to find them is to start in the middle and work outwards. Imagine a spacecraft sitting at the Earth/Moon L1 point, between the two bodies. If it's given a tiny push, it starts to run 'downhill' as it loses potential energy and gains kinetic. Some pushes send it Earthwards, eventually orbiting our home planet. Others send it Moonwards into a lunar capture orbit. By reversing the path from L1 to the Earth, and tagging on a suitable path from L1 to the Moon, we get a highly efficient trajectory from the Earth to the Moon with a junction at L1.

As it happens, L1 is a great place to make small course changes. The natural dynamics of the spacecraft near L1 are chaotic, so very small changes in position or speed create large changes to the trajectory. By exploiting chaos, we can redirect our spacecraft to other destinations in a fuel-efficient, though possibly slow, manner.

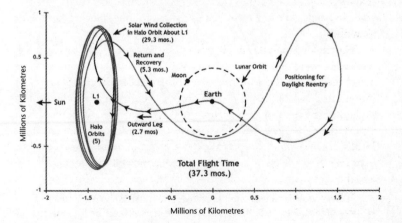

Trajectory of the Genesis mission.

The tube trick was first used in 1985 to redirect the almost-dead International Sun–Earth Explorer *ISEE-3* to rendezvous with comet Giacobini–Zinner. In 1990 Belbruno contacted the Japanese space agency about one of their probes, *Hiten*, which had completed its main mission and was low on fuel. He presented a trajectory that would temporarily park it in lunar orbit, and then redirect it to the L4 and L5 points to look for trapped dust particles. The same trick was used again for the *Genesis* mission to bring back samples of particles in the solar wind.[5]

Mathematicians and engineers who wanted to repeat that trick, and to find others of the same kind, tried to understand what really made it work. They homed in on those special places in the energy landscape, analogous to mountain passes. These create 'bottlenecks' that would-be travellers must navigate. There are specific 'inbound' and 'outbound' paths, analogous to the natural routes through a pass. To follow these inbound and outbound paths exactly, you have to travel at just the right speed. But if your speed is slightly different, you can still stay near those paths. To plan an efficient mission profile, you decide which tubes are relevant. You route the spacecraft along the first inbound tube, and when it gets to the Lagrange point a quick burst on the motors redirects it along an outbound tube. This flows into another inbound tube … and so it goes.

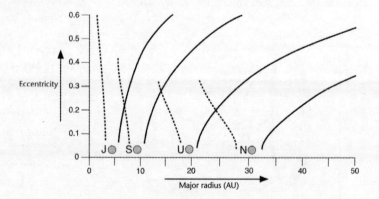

Low-energy paths linked to L1 (dotted) and L2 (solid) points for the four outer planets (J, S, U, N). Those for the inner planets are too small to be visible on this scale. Intersections, where the surrounding tubes meet, provide switching points for low-energy transfer orbits.

In 2000 Wang Sang Koon, Martin Lo, Jerrold Marsden, and Shane Ross used tubes to design a tour of the moons of Jupiter, with a gravitational boost near Ganymede followed by a tube trip to Europa. A more complex route, requiring even less energy, includes Callisto as well. It uses five-body dynamics: Jupiter, the three moons, and the spacecraft.

In 2002 Lo and Ross computed natural paths in the energy landscape, leading into and away from the L1 and L2 points of the planets of the solar system, and found that they intersect each other. The picture illustrates these paths in a Poincaré section. The dotted curve emanating from Saturn (S) crosses the solid curve emanating from Jupiter (J), giving a low-energy transfer orbit between the two planets concerned. The same goes for the other intersections. So, starting at Neptune, a spacecraft can transfer efficiently to Uranus, then Saturn, then Jupiter, switching between the L1 and L2 points at each planet. The same process can be continued into the inner solar system, or reversed to move outwards, step by step. This is the mathematical skeleton of the interplanetary superhighway.

In 2005, Michael Dellnitz, Oliver Junge, Marcus Post, and Bianca Thiere used tubes to plan an energy-efficient mission from the Earth to Venus. The main tube links the Sun/Earth L1 point to the Sun/Venus L2 point. As a comparison, this route uses only one third of the fuel required by the European Space Agency's *Venus Express* mission, because it can use low-thrust engines; the price is a lengthening of the transit time from 150 days to about 650 days.

The influence of tubes may go further. Dellnitz has discovered a natural system of tubes connecting Jupiter to each of the inner planets. This is a hint that Jupiter, the dominant planet of the solar system, plays the role of a celestial Grand Central Station. Its tubes may well have organised the formation of the entire solar system, determining the spacings of the inner planets. It doesn't explain, or even support, the Titius-Bode law; instead, it shows that the true organisation of planetary systems comes from the subtle patterns of nonlinear dynamics.

Great Balls of Fire

> We may determine the forms of planets, their distances, their sizes, and their motions – but we can never know anything of their chemical composition.
>
> Auguste Comte, *The Positive Philosophy*

WITH TWENTY-TWENTY HINDSIGHT it's easy to poke fun at poor Comte, but in 1835 it was inconceivable that we could ever find out what a planet is made of, let alone a star. The quote says 'planets', but elsewhere he stated that it would be even harder to find the chemistry of a star. His main point was that there are limits to what science can discover.

As so often happens when scholars of high repute declare something to be impossible, Comte's deeper point was correct, but he chose precisely the wrong example. Ironically, the chemical composition of a star, even one that's thousands of light years away, is now one of its easiest features to observe. Provided you don't want too much detail, the same goes for galaxies, millions of light years away. We can even learn a lot about the atmospheres of planets, which shine by reflected starlight.

Stars raise lots of questions, in addition to what they're made of. What are they, how do they shine, how do they evolve, how far away are they? By combining observations with mathematical models, scientists have inferred detailed answers to all of these questions, even though visiting a star with today's technology is virtually impossible. Let alone tunnelling inside one.

✦

The discovery that rubbished Comte's example was an accident. Joseph Fraunhofer started out as a glassmaker's apprentice, and was nearly killed when his workshop collapsed. Maximilian IV Joseph, prince-elector of Bavaria, organised a rescue, took a shine to the young man, and financed his education. Fraunhofer became an expert in the production of glass for optical instruments, eventually becoming Director of the Optical Institute in Benediktbeuern. He built high-quality telescopes and microscopes, but his most influential scientific breakthrough came in 1814 when he invented a new instrument – the spectroscope.

Newton worked on optics as well as mechanics and gravity, and he discovered that a prism splits white light into its component colours. Another way to split light is to use a diffraction grating, a flat surface with closely spaced ruled lines. Light waves reflected from the grating interfere with each other. The geometry of the waves implies that light of a given wavelength (or frequency, which is lightspeed divided by wavelength) is reflected most strongly at specific angles. There, the wave peaks coincide, so they reinforce each other. In contrast, almost no light is reflected at angles where the waves interfere destructively, so that the peak of one wave meets the trough of another. Fraunhofer combined a prism, a diffraction grating, and a telescope to create an instrument that can split light into its components and measure their wavelengths to high accuracy.

One of his first discoveries was that the light emitted by a fire has a characteristic orange hue. Wondering whether the Sun is basically a ball of fire, he pointed his spectroscope at it to look for light of that wavelength. Instead, he observed a full spectrum of colours, as Newton had done, but his instrument was so precise that it also revealed the presence of mysterious dark lines at numerous wavelengths. Earlier, William Wollaston had noticed about six of these lines; Fraunhofer eventually found 574.

By 1859 the physicist Gustav Kirchhoff and the chemist Robert Bunsen, famous for his burner, had demonstrated that these lines appear because atoms of various elements absorb light of specific wavelengths. The Bunsen burner was invented to measure these wavelengths in the laboratory. If you know, say, the wavelengths that potassium produces, and you find a corresponding line in the Sun's spectrum, the Sun must contain potassium. Fraunhofer applied this idea to Sirius, thereby observing the first stellar spectrum. Looking at other stars, he noticed that they had different spectra. The implication was huge: not only can we find out what stars are made of, but different stars are made of different things.

A new branch of astronomy, stellar spectroscopy, was born.

Two main mechanisms create spectral lines. Atoms can absorb light of a given wavelength, creating an absorption line, or they can emit it, creating an emission line. The characteristic yellowish light of sodium street lamps is an emission line of sodium. Working sometimes together and sometimes separately, Kirchhoff and Bunsen used their technique to discover two new elements, caesium and rubidium. Soon two astronomers, Jules Janssen and Norman Lockyer, would go one better, discovering an element that – at that time – had never been detected on Earth.

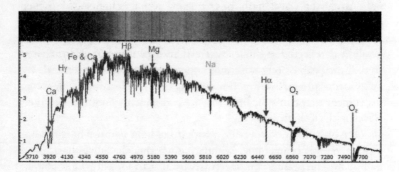

Spectrum of a star. Top: as seen in a spectroscope. Bottom: brightness at each wavelength. Absorption lines shown are (left to right) calcium, hydrogen gamma, iron and calcium, hydrogen beta, magnesium, sodium, hydrogen alpha, oxygen, oxygen.

✦

In 1868 Janssen was in India, observing a solar eclipse, hoping to find the chemistry of the Sun's chromosphere. This is a layer in the Sun's atmosphere that lies immediately above the visible layer, which is the photosphere. The chromosphere is so faint that it can be observed only during a total eclipse, when it has a reddish hue. The photosphere creates absorption lines, but the chromosphere creates emission lines. Janssen found a very strong bright yellow emission line (which therefore came from the chromosphere) with a wavelength of 587·49 nanometres, and thought it corresponded to sodium. Soon after, Lockyer named it the D_3 spectral line, because sodium has two lines at similar wavelengths named D_1 and D_2. However, it has no line at the D_3 wavelength, so that line isn't a sign of sodium.

In fact, no known atom had such a line. Lockyer realised they'd stumbled across a previously unknown element. He and the chemist Edward Frankland named it helium, after the Greek word *helios* for 'sun'. By 1882 Luigi Palmieri had spotted the D_3 line on Earth in a sample of volcanic lava from Mount Vesuvius. Seven years later William Ramsay obtained samples of helium by applying acid to a mineral called cleveite, which contains uranium along with several 'rare earth' elements. Helium turned out to be a gas at room temperature.

So far this story is mostly chemistry, aside from the mathematical theory of diffraction. But the tale now took an unexpected twist, bringing it into the highly mathematical realm of particle physics. In 1907 Ernest Rutherford and Thomas Royds were studying alpha particles, emitted by radioactive materials. To find out what they were, they trapped them in a glass tube containing – nothing. A vacuum. The particles passed through the wall of the tube, but lost energy and couldn't get out again. The spectrum of the contents of the tube had a strong D_3 line. Alpha particles are the atomic nuclei of helium.

To cut a long story short, the combined efforts of these scientists led to the discovery of the second commonest element in the universe, after hydrogen. But helium isn't common *here*. We get most of it by distilling natural gas. It has numerous important scientific uses: meteorological balloons, low-temperature physics, magnetic resonance imaging scanners for medical use. And, potentially, as the main fuel of fusion reactors, a cheap and relatively safe form of energy if anyone can make the things work. So what do we squander this vital element on? Toy balloons for children's parties.

Most of the helium in the universe occurs in stars and interstellar gas clouds. That's because it was originally made in the early stages of the Big Bang, and it's also the main outcome of fusion reactions in stars. We see it in the Sun because the Sun isn't just made of it, along with a lot of hydrogen and lots of other elements in lesser amounts: it *makes* it … from hydrogen.

A hydrogen atom consists of one proton and one electron. A helium atom has two protons, two neutrons, and two electrons; an alpha particle omits the electrons. In a star, the electrons are stripped away and the reactions involve just the nucleus of the atom. In the core of the Sun, where the temperature is 14 million kelvin, four hydrogen nuclei – four protons – are squashed together by massive gravitational forces. They merge to create an alpha particle, two positrons, two neutrinos, and a lot of energy. The positrons and neutrinos allow two protons to become two neutrons. At a deeper level we should really consider the constituent quarks, but this description suffices. A similar reaction makes a 'hydrogen bomb' explode with devastating force, thanks to that energy release, but it involves other isotopes of hydrogen: deuterium and tritium.

✦

The early stages of a new branch of science are much like butterfly collecting: catch everything you can and try to arrange your specimens in a rational manner. Spectroscopists collected stellar spectra, and classified the stars accordingly. In 1866 Angelo Secchi used spectra to group stars into three distinct classes, corresponding roughly to their predominant colours: white and blue, yellow, orange and red. Later he added two more classes.

Around 1880 Pickering began a survey of stellar spectra, published in 1890. Williamina Fleming performed most of the subsequent classification, using a refinement of Secchi's system symbolised by letters of the alphabet from A to Q. After a complicated series of revisions, the current Morgan–Keenan system emerged, which uses O, B, A, F, G, K, and M. Stars of type O have the hottest surface temperature, type M the coolest. Each class splits into smaller ones numbered 0–9, so that the temperature goes down as the number goes up. Another key variable is the luminosity of a star – its intrinsic 'brightness' at all wavelengths, measured as the total radiation it emits every second.[1] Stars are also given a luminosity class, mostly written as a Roman numeral, so this scheme has two parameters, corresponding roughly to temperature and luminosity.

   Class O stars, for instance, have a surface temperature in excess of 30,000 K, appear blue to the eye, have mass at least 16 times that of the Sun, show weak hydrogen lines, and are very rare. Class G stars have a surface temperature between 5200 and 6000 K, are pale yellow, have mass between 0·8 and 1·04 times that of the Sun, show weak hydrogen lines, and constitute about 8% of all known stars. Among them is our Sun, type G2. Class M stars have a surface temperature between 2400 and 3700 K, are orangey-red, have mass between 0·08 and 0·45 times that of the Sun, show very weak hydrogen lines, and constitute about 76% of all known stars.

   The luminosity of a star correlates with its size, and the different luminosity classes have names ranging from hypergiants, through super-giants, giants, subgiants, dwarfs (or main sequence), and subdwarfs. So a particular star may be described as a blue giant, a red dwarf, and so on.

   If you plot the temperature and luminosity of stars on a graph, you don't get a random scatter of dots. You get a shape like a backward Z. This is the Hertzsprung–Russell diagram, introduced around 1910 by Ejnar Hertzsprung and Henry Russell. The most prominent features are a cluster of bright, coolish giants and supergiants at top right, a curvy diagonal 'main sequence' from hot and bright to cool and dim, and a sparse cluster of hot, dim white dwarfs at bottom left.

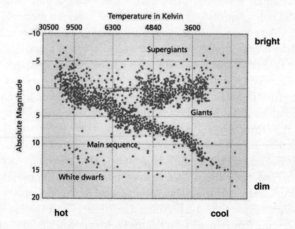

Hertzsprung–Russell diagram. Absolute magnitude is related to luminosity, with −10 being very bright and +20 very dim.

The study of stellar spectra went beyond butterfly collecting when scientists started using them to work out how stars produce their light and other radiation. They quickly realised that a star isn't just a gigantic bonfire. If its energy source had been ordinary chemical reactions, the Sun would have burnt to a cinder long ago. The Hertzsprung–Russell diagram also suggested that stars might evolve along the backward Z from top right to bottom left. It seemed sensible: they'd be born as giants, shrink to dwarfs, and progress down the main sequence to become subdwarfs. As they shrunk, they would convert gravitational energy into radiation, a process called the Kelvin–Helmholtz mechanism. From this theory, astronomers of the 1920s deduced that the Sun is about 10 million years old, attracting the ire of geologists and evolutionary biologists who were convinced it's far older.

It wasn't until the 1930s that the astronomers caved in, having realised that stars get most of their energy from nuclear reactions, not gravitational collapse, and the proposed evolutionary pathway was wrong. A new area of science, astrophysics, was born. It employs sophisticated mathematical models to analyse the dynamics and evolution of stars, from the moment of their birth to the instant of their death. The main ingredients for these models come from nuclear physics and thermodynamics.

In Chapter 1 we saw how stars form when a vast primordial gas cloud collapses under its own gravity. There, we focused on the dynamics, but nuclear reactions add new detail. The collapse releases gravitational energy, which heats the gas to create a protostar, a very hot spheroid of spinning gas. Its main constituent is hydrogen. If the temperature reaches 10 million kelvin, hydrogen nuclei – protons – begin to fuse together, producing deuterium and helium. Protostars with initial mass less than 0·08 that of the Sun never get that hot, and fizzle out to form brown dwarfs. They shine dimly, mostly by the light of deuterium fusion, and fade away.

Stars hot enough to light up start out using the proton–proton chain reaction. First, two protons fuse to form a diproton (a lightweight form of helium) and a photon. Then one proton in the diproton emits a positron and a neutrino and becomes a neutron; now we have a nucleus of deuterium. This step, though relatively slow, releases a small amount of energy. The positron collides with an electron and they mutually annihilate to create two photons and a bit more energy. After about four seconds the deuterium nucleus fuses with another proton to make an isotope of helium, helium-3; quite a lot more energy is released.

At this stage there are three options. The main one fuses two helium-3 nuclei to create ordinary helium-4, plus two hydrogen nuclei and even more energy. The Sun uses this pathway 86% of the time. A second option creates nuclei of beryllium, which changes to lithium, which fuses with hydrogen to make helium. Various particles are also given off. The Sun uses this pathway 14% of the time. The third pathway involves nuclei of beryllium and boron, and occurs 0·11% of the time in the Sun. Theoretically there's a fourth option where helium-3 fuses with hydrogen to go directly to helium-4, but it's so rare that it's never been observed.

Astrophysicists represent these reactions by equations like

$$_1^2 D + {}_1^2 H \rightarrow {}_2^3 He + \gamma + 5.49 \text{ MeV}$$

where D = deuterium, H = hydrogen, He = helium, the top index is the number of neutrons, the bottom one is the number of protons, γ is a photon, and MeV is a unit of energy (megaelectronvolt). I mention this not because I want you to follow the process in detail, but to show that it *can* be followed in detail and it has a definite mathematical structure.

Earlier I mentioned the theory that stars evolve, so that their characteristic combination of temperature and luminosity moves across the Hertzsprung–Russell diagram. There's some merit to the idea, but the original details were wrong, and different stars follow different paths – in roughly the opposite direction to what was first thought.[2] When a star has come into being, it takes its place somewhere in the main sequence of the Hertzsprung–Russell diagram. The location depends on the mass of the star, which determines its luminosity and temperature. The main forces affecting the dynamics of the star are gravity, which makes it contract, and radiation pressure caused by hydrogen fusion, which makes it expand. A stable feedback cycle plays these forces off against each other to reach an equilibrium state. If gravity started to win, the star would contract, heating it up and increasing the radiation levels to restore the balance. Conversely, if radiation started to win, the star would expand, cool, and gravity would bring it back to equilibrium.

This balancing act continues until the fuel starts to run out. This takes hundreds of billions of years for slow-burning red dwarfs, 10 billion years or so for stars like the Sun, and a few million years for hot, massive O-type stars. At that point, gravity takes over and the star's core contracts. Either the core becomes hot enough to initiate helium fusion, or the core turns into degenerate matter – a kind of atomic gridlock – arresting the gravitational collapse. The mass of the star determines which occurs; a few cases exemplify the variety.

If the star's mass is less than a tenth of the Sun's, it remains on the main sequence for 6–12 trillion years, eventually becoming a white dwarf.

A star with the mass of the Sun develops an inert helium core surrounded by a hydrogen-burning shell. This causes the star to expand, and as its outer layers cool, it becomes a red giant. The core collapses until its matter becomes degenerate. The collapse releases energy, heating the surrounding layers, which start to transport heat by convection rather than just radiating it. The gases become turbulent and flow from the core towards the surface, and back again. After a billion years or thereabouts, the degenerate helium core becomes so hot that helium nuclei fuse to make carbon, with beryllium as a short-lived intermediary. Depending on other factors, the star may then evolve further to an asymptotic giant. Some stars of this kind pulsate, alternately expanding and shrinking, and their temperature also oscillates. Eventually, the star cools and becomes a white dwarf.

The Sun has about 5 billion years left before it becomes a red giant. At that point Mercury and Venus will be engulfed as the Sun expands. At that stage the Earth will probably orbit just outside the surface of the Sun, but tidal forces and friction with the chromosphere will slow it down. Eventually it, too, will be swallowed up. This won't affect the long-term future of the human race, because the average lifetime of a species is only a few million years.

A sufficiently massive star, much larger than the Sun, starts fusing helium before the core becomes degenerate, and explodes to form a supernova. A star with more than 40 times the mass of the Sun blows away much of its matter through radiation pressure, stays very hot, and embarks on a series of stages at which the main element at the core is replaced by one further up the periodic table. The core becomes assorted into concentric layers: iron, silicon, oxygen, neon, carbon, helium, hydrogen. The core of the star can end up as a white dwarf or a black dwarf, which is a white dwarf that has lost so much energy that it stops shining. Alternatively, a sufficiently massive degenerate core can form a neutron star, or in more extreme cases, a black hole: see Chapter 14.

Again, the details don't matter, and I've greatly simplified a very complex branching tree of possible evolutionary histories. The mathematical models used by astrophysicists govern the range of possibilities, the order in which they arise, and the conditions that lead to them. The rich variety of stars, of all sizes, temperatures, and colours, has a common origin: nuclear fusion starting from hydrogen, subject to the competing forces of radiation pressure and gravity.

A common thread running through the story is how fusion converts simple hydrogen nuclei into more complex nuclei: helium, beryllium, lithium, boron, and so on.

And that leads to another reason why stars are important.

✦

'We are stardust,' sang Joni Mitchell. It's a cliché, but clichés are often true. Earlier, Arthur Eddington had said the same in the *New York Times Magazine*: 'We are bits of stellar matter that got cold by accident, bits of a star gone wrong.' Try setting *that* to music.

According to the Big Bang, the only (nucleus of an) element in the early universe was hydrogen. Between 10 seconds and 20 minutes after the universe first appeared, Big Bang nucleosynthesis using reactions like those just described created helium-4, plus tiny amounts of deuterium, helium-3, and lithium-7. Short-lived radioactive tritium and beryllium-7 also arose, but quickly decayed.

The hydrogen alone was enough to create gas clouds, which collapsed to make protostars, and then stars. More elements were born in the nuclear maelstrom inside stars. In 1920 Eddington suggested that stars are powered by fusion of hydrogen into helium. In 1939 Hans Bethe studied the proton–proton chain and other nuclear reactions in stars, giving flesh to the bones of Eddington's theory. In the early 1940s George Gamow argued that almost all elements came into being during the Big Bang.

In 1946 Fred Hoyle suggested that the source of everything above hydrogen was not the Big Bang as such, but subsequent nuclear reactions inside stars. He published a lengthy analysis of reaction routes leading to all elements up to iron.[3] The older the galaxy, the richer its brew of elements. In 1957 Margaret and Geoffrey Burbidge, William Fowler, and Hoyle published 'Synthesis of the elements in stars'.[4] Generally referred to as B[2]FH, this famous paper founded the theory of stellar nucleosynthesis – a fancy way to say the title – by sorting out many of the most important nuclear reaction processes. Soon astrophysicists had a convincing account, predicting proportions of elements in the Galaxy that (mostly) match what we observe.

At that time the story stopped at iron, because it's the most massive nucleus that can arise through the silicon-burning process, a chain of reactions starting from silicon. Repeated fusion with helium leads to calcium, and thence through a series of unstable isotopes of titanium, chromium, iron, and nickel. This isotope, nickel-56, constitutes a barrier to further progress, because another helium-fusion step would use up energy instead of producing it. The nickel isotope decays to radioactive cobalt-56, which turns into stable iron-56.

To get beyond iron, the universe had to invent another trick.

Supernovas.

A supernova is an exploding star. A nova is a less energetic form, which would lead us off topic. Kepler saw a nova in 1604, the last seen happening in the Galaxy, though remains of two more recent ones have been spotted. Basically, a supernova is an extreme version of a nuclear bomb, and when it happens the star outshines a galaxy. It gives off as much radiation as the Sun will over its entire lifetime. There are two causes. A white dwarf can have extra matter dumped into it by gobbling up a companion star, which makes it hotter and ignites carbon fusion; this 'runs away' unchecked and the star blows up. Alternatively, the core of a very massive star can collapse, and the energy released can trigger such an explosion.

Either way, the star comes to pieces at one tenth of the speed of light, creating a shockwave. This collects gas and dust, forming a growing shell, a supernova remnant. And that's how elements higher up the periodic table than iron came to be, and how they spread across galactic distances.

I said that the predicted proportions of elements *mostly* match observations. A glaring exception is lithium: the actual amount of lithium-7 is only a third of what theory predicts, while there's about a thousand times too much lithium-6. Some scientists think this is a

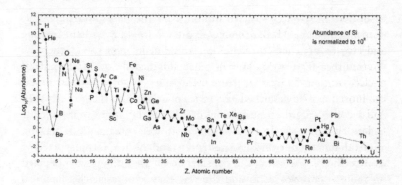

Estimated abundances of chemical elements in the solar system. The vertical scale is logarithmic, so fluctuations are much larger than they appear to be.

minor error that can probably be fixed up by finding new pathways or new scenarios for lithium formation. Others see it as a serious problem, possibly requiring new physics going beyond the standard Big Bang.

There's a third possibility: more lithium-7 is present, but not where we can detect it. In 2006 Andreas Korn and coworkers reported that the lithium abundance in the globular cluster NGC 6397, in the general region of the Large Magellanic Cloud, is much as predicted by Big Bang nucleosynthesis.[5] They suggest that the apparent lack of lithium-7 in stars of the Galaxy's halo – about one quarter of the prediction – may be a sign that these stars may appear to have lost lithium-7 because turbulent convection has transported it to deeper layers, where it can no longer be detected.

The response to the lithium discrepancy raises a potential issue with the Big Bang nucleosynthesis predictions. Suppose you're calculating the abundance of various elements. The commonest nuclear reactions probably account for much of what happened, leading to values not too far removed from reality in most cases. Now you start working on the discrepancies. Too little sulphur? Hmm, let's find new routes to sulphur. When we do, and the numbers look right, sulphur's sorted, and we go on to zinc. What we don't do is keep looking for even more routes to sulphur. I'm not suggesting anyone deliberately does this kind of thing, but selective reporting like this is natural, and it's happened elsewhere in science. Perhaps lithium isn't the only discrepancy. By focusing on cases where proportions are too small, we may be missing ones where an extended calculation would make them too large.

Another feature of stars that depends heavily on mathematical models is their detailed structure. Most stars, at a given stage in their evolution, can be described as a series of concentric shells. Each shell has its own specific composition and 'burns' by appropriate nuclear reactions. Some shells are transparent to electromagnetic radiation, and radiate heat outwards. Some are not, and the heat is transported by convection. These structural considerations are intimately connected to the evolution of stars and the way they synthesise the chemical elements.

✦

Sorting out a proportion that was too small led Hoyle to a famous prediction. When he did the sums for the abundance of carbon, there wasn't enough of it. However, *we* exist, with carbon as a vital ingredient. Since we're stardust, the stars must somehow make a lot more carbon than Hoyle's sums indicated. So he predicted the existence of a hitherto unknown resonance in the carbon nucleus, which would make it much easier to form carbon.[6] The resonance was then observed, roughly where Hoyle had predicted. It's often presented as a triumph for the anthropic principle: our existence imposes constraints on the universe.

A critical analysis of this story rests on some nuclear physics. The natural route to carbon is the triple-alpha process, occurring in a red giant star. Helium-4 has two protons and two neutrons. The main isotope of carbon has six of each. So three nuclei of helium (alpha particles) can fuse to make carbon. Great, but... Two helium nuclei frequently collide, but if we want carbon, the third has to smash into them just as they do so. A triple collision in a star is terribly rare, so carbon can't arise by that route. Instead, two helium nuclei fuse to make beryllium-8; then a third helium nucleus fuses with the result to make carbon. Unfortunately beryllium-8 decays after 10^{-16} seconds, giving the helium nucleus a very small target. This two-step method can't make enough carbon.

Unless ... the energy of carbon is very close to the combined energies of beryllium-8 and helium. This is a nuclear resonance, and it led Hoyle to predict a then unknown state of carbon at an energy 7·6 MeV above the lowest energy state. A few years later a state with energy 7·6549 MeV was discovered. But the energies of beryllium-8 and helium add to 7·3667 MeV, so carbon's newly discovered state has a bit too much energy.

Where does it come from? It's almost exactly the energy supplied by the temperature of a red giant.

This is one of the favourite examples of proponents of 'fine-tuning', the idea that the universe is exquisitely finely tuned for life to exist. I'll come back to that in Chapter 19. Their point is that, without carbon, we wouldn't be here. But that much carbon requires fine-tuning a star and a nuclear resonance, and those depend on fundamental physics. Later, Hoyle expanded on this idea:[7]

> Some super-calculating intellect must have designed the properties of the carbon atom, otherwise the chance of my finding such an atom through the blind forces of nature would be utterly minuscule. A common sense interpretation of the facts suggests that a superintellect has monkeyed with physics, as well as with chemistry and biology, and that there are no blind forces worth speaking about in nature.

It sounds remarkable, and it surely can't be coincidence. Indeed, it's not. But the reason debunks the fine-tuning. Every star has its own thermostat, a negative feedback loop in which the temperature and the reaction adjust each other so that they fit. The 'fine-tuned' resonance in the triple-alpha process is no more amazing than a coal-burning fire being at just the right temperature to burn coal. That's what coal fires do. In fact, it's no more amazing than our legs being just the right length to reach the ground. That's a feedback loop too: muscles and gravity.

It was a bit naughty of Hoyle to phrase his prediction in terms of human existence. The real point is that the *universe* has too little carbon. Of course, it's still amazing that red giants and atomic nuclei exist at all, that they manufacture carbon from hydrogen, and that some carbon is eventually incorporated into us. But those are different issues. The universe is endlessly rich and complex, and all sorts of wonderful things happen. But we shouldn't confuse outcomes with causes, and imagine that the purpose of the universe is to make humans.

One reason I've mentioned this (aside from a distaste for exaggerated fine-tuning claims) is that the whole story has been made irrelevant by the discovery of a new way for stars to make carbon. In 2001 Eric Feigelson and coworkers discovered 31 young stars in the Orion nebula. They're all much the same size as the Sun, but they're extremely active, sending out X-ray flares a hundred times more powerful than solar flares today, a hundred times as often. Protons in those flares have enough energy to create all sorts of heavy elements in a dust disc around the star. So you don't need a supernova to get them. That suggests we need to revise the calculations about the origin of chemical elements, including carbon. Effects that look impossible may just stem from a lack of human imagination. Proportions that look right might change if we thought a bit harder.

✦

To Greek philosophers, the Sun was a perfect embodiment of celestial geometry, an unblemished sphere. But when ancient Chinese astronomers viewed the Sun through a haze, they saw it was spotty. Kepler noticed a spot on the Sun in 1607, but thought it was Mercury in transit. In 1611 Johannes Fabricius published *Maculis in Sole Observatis, et Apparente earum cum Sole Conversione Narratio* (Narration on Spots Observed on the Sun and their Apparent Rotation with the Sun), whose title is self-explanatory. In 1612 Galileo observed irregular dark spots on the Sun and made drawings showing that they move, confirming Fabricius's contention that the Sun rotates. The existence of sunspots demolished the longstanding belief in the Sun's perfection, and sparked a heated priority dispute.

The number of sunspots varies from year to year, but there's a fairly regular pattern, an 11-year cycle ranging from hardly any spots to a hundred or more per year. Between 1645 and 1715 the pattern was broken, with hardly any sunspots being seen. This period is called the Maunder minimum.

There may be a link between sunspot activity and climate, but if so it's probably weak. The Maunder minimum coincided with the middle of the Little Ice Age, a prolonged period of unusually cold temperatures in Europe. So did a previous period of low sunspot activity, the Dalton minimum (1790–1830), which includes the famous 'year without a summer' of 1816, but the low temperatures that year resulted from a huge volcanic explosion, Mount Tambora in Sumbawa, Indonesia. The Little Ice Age may also have been caused, at least in part, by high levels of volcanic activity.[8] The Spörer minimum (1460–1550) is associated with another cooling period; the evidence for it comes from the proportion of the isotope carbon-14 in tree rings, which is associated with solar activity. Sunspot records as such were not being kept at that time.

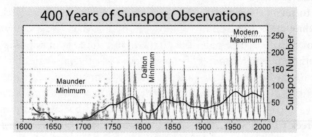

Changes in sunspot numbers.

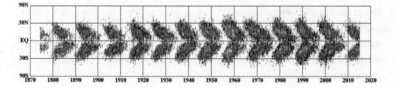

Sunspots plotted by latitude.

Plotting the latitude of sunspots as well as their number reveals a curious pattern like a series of butterflies. The cycle begins with spots near the pole, which gradually appear nearer the equator as the numbers approach their maximum. In 1908 George Hale took the first step towards understanding this behaviour when he linked sunspots to the Sun's magnetic field, which is immensely strong. Horace Babcock modelled the dynamics of the Sun's magnetic field in its outermost layers, relating the sunspot cycle to periodic reversals of the solar dynamo.[9] In his theory the full cycle lasts 22 years, with a north/south reversal of the field separating the two halves.

Sunspots appear dark only in contrast to their surroundings; a sunspot's temperature is about 4000 K and the gases around them are at 5800 K. They are analogous to magnetic storms in the superheated solar plasma. Their mathematics is governed by magnetohydrodynamics, the study of magnetic plasmas, which is highly complex. They seem to be the top ends of tubes of magnetic flux, originating deep in the Sun.

The general form of the Sun's magnetic field is a dipole, like a bar magnet, with a north pole, a south pole, and field lines streaming from one to the other. The poles are aligned with the axis of rotation, and during the normal running of the sunspot cycle the polarities flip every 11 years. So the magnetic pole in the Sun's 'northern hemisphere' is sometimes magnetic north, and at other times magnetic south. Sunspots tend to appear in linked pairs, with a field like a bar magnet pointing east–west. The first to appear has the same polarity as the nearest pole of the main magnetic field, the second, which trails behind, has the opposite polarity.

The solar dynamo, which drives its magnetic field, is caused by convective cyclones in the outer 200,000 km of the Sun, in conjunction with how the star spins: faster at the equator than near its poles. Magnetic fields in a plasma are 'trapped' and tend to move with it, so the initial positions of the field lines, looping between the poles at right angles to the equator, start to wind up as the equatorial region pulls them ahead of the polar ones. This twists the field lines, entwining fields of opposite polarity. As the Sun rotates, the magnetic field lines become ever more tightly wound, and when the stresses reach a critical value, the tubes curl up and hit the surface. The field lines stretch and the associated sunspots drift towards the pole. The trailing spot gets to the pole first, and since it has the opposite polarity, it – helped by numerous similar events – causes the Sun's magnetic field to flip. The cycle repeats with a reversed field.

One theory of the Maunder minimum is that the Sun's dipole field is supplemented by a quadrupole field – like two bar magnets side by side.[10] If the period of reversal of the quadrupole is slightly different from that of the dipole, the two 'beat' like two musical notes that are close together but not identical. The result is a long-period oscillation in the average size of the field during a cycle, and when it dies down few sunspots appear anywhere. Furthermore, a quadrupole field has opposite polarities in the two hemispheres, so it reinforces the dipole field in one hemisphere while cancelling it in the other. So the few sunspots that do appear all arise in the same hemisphere, which is what happened during the Maunder minimum. Similar effects have been observed indirectly in other stars, which can also have sunspots.

Combined dipole and quadrupole fields in a simple model of the solar dynamo, in which the total energy oscillates while the amplitude repeatedly dies down and increases again.

Field lines that poke above the photosphere can create solar prominences, huge loops of hot gas. These loops can break and reconnect, allowing plasma and magnetic field lines to be blown away on the solar wind. This causes solar flares, which can disrupt communications and damage electrical power grids and artificial satellites. Often these are followed by coronal mass ejections, where huge amounts of matter are released from the corona, a tenuous region outside the photosphere visible to the eye during an eclipse of the Sun.

✦

A basic question is: how far away are the stars? As it happens, the only reason we know the answer, for anything more than a few dozen light years away, also depends on astrophysics, although initially the key observation was empirical. Henrietta Leavitt found a standard candle, and put a yardstick to the stars.

In the sixth century BC the ancient Greek philosopher and mathematician Thales estimated the height of an Egyptian pyramid using geometry, by measuring its shadow and his own. The ratio of the height of the pyramid to the length of its shadow is the same as the ratio of Thales's height to the length of *his* shadow. Three of those lengths can easily be measured, so you can work out the fourth. His ingenious method is a simple example of what we now call trigonometry. The geometry of triangles relates their angles to their sides. Arab astronomers developed the idea for instrument-making, and it came back down to Earth in medieval Spain, for surveying. Distances are hard to measure, because there are often obstacles in the way, but angles are easy. You can use a pole and some string, or better a telescopic sight, to measure the direction of a distant object. You start by measuring, very accurately, a known baseline. Then you measure the angles from either end to some other point, and calculate the distances to that point. Now you have two more known lengths, so you can repeat the process, 'triangulating' the area you wish to map, and calculating all distances from just that one measured baseline.

Eratosthenes famously used geometry to calculate the size of the Earth by looking down a well. He compared the angle of the noonday Sun at Alexandria and Syene (present-day Aswan) and worked out the distance between them from the time it took camels to travel from one to the other. Knowing the size of the Earth, you can observe the Moon from two different locations and deduce the distance to the Moon. Then you can use that to find the distance to the Sun.

How? Around 150 BC, Hipparchus realised that when the Moon's phase is exactly half-full, the line from the Moon to the Sun is perpendicular to the line from the Earth to the Moon. Measure the angle between this baseline and the line from the Earth to the Sun, and you can calculate how far away the Sun is. His estimate, 3 million kilometres, was far too small: the correct value is 150 million. He got it wrong because he thought the angle was 87 degrees when actually it's very close to a right angle. With better equipment, you can get an accurate estimate.

This bootstrapping process goes one step further. You can use the orbit of the Earth as a baseline to find the distance to a star. The Earth goes halfway round its orbit in six months. Astronomers define the *parallax* of a star to be half the angle between the two sight lines to the star, observed from opposite ends of the Earth's orbit. The star's distance is approximately proportional to the parallax, and a parallax of one second of arc corresponds to about 3·26 light years. This unit is the parsec (*par*allax arc*sec*ond), and many astronomers prefer it to the light year for this reason.

James Bradley tried to measure the parallax of a star in 1729, but his apparatus wasn't accurate enough. In 1838 Friedrich Bessel used one of Fraunhofer's heliometers, a sensitive new design of telescope delivered after Fraunhofer's death, to observe the star 61 Cygni. He measured a parallax of 0·77 arcseconds, comparable to the width of a tennis ball ten kilometres away, giving a distance of 11·4 light years, very close to the current value. That's 100 trillion kilometres, demonstrating how puny our world is compared to the universe that surrounds it.

The diminution of humanity was not yet finished. Most stars, even in our own Galaxy, show no measurable parallax, implying that they are much further away than 61 Cygni. But when there's no detectable parallax, triangulation breaks down. Space probes could provide a longer baseline, but not orders of magnitude longer, which is what's needed for distant stars and galaxies. Astronomers needed to think of something radically different to continue their climb up the cosmic distance ladder.

✦

You have to work with what's possible. One readily observed feature of a star is its apparent brightness. This depends on two factors: how bright it really is – its intrinsic brightness or luminosity – and how far away it is. Brightness is like gravity, falling off as the inverse square of the distance. If a star with the same intrinsic brightness as 61 Cygni has an apparent brightness one ninth of that amount, it must be three times as far away.

Unfortunately, intrinsic brightness depends on the type of star, its size, and precisely which nuclear reactions are happening inside it. To make the apparent brightness method work we need a 'standard candle' – a type of star whose intrinsic brightness is known, or can be inferred *without* knowing how far away it is. That's where Leavitt came in. In the 1920s Pickering hired her as a human 'computer', carrying out the repetitive task of measuring and cataloguing the luminosities of stars in the Harvard College Observatory's collection of photographic plates.

Most stars have the same apparent brightness all the time, but some, which naturally arouse special interest among astronomers, are variable: their apparent brightness increases and decreases in a regular periodic pattern. Leavitt made a special study of variable stars. There are two main reasons for their variability. Many stars are binary – two stars orbit round a common centre of mass. If it so happens that the Earth lies in the plane of these orbits, then the stars pass in front of each other at regularly spaced times. When they do, the result is just like – indeed, *is* – an eclipse: one star gets in the way of the other and temporarily blocks its light output. These 'eclipsing binaries' are variable stars, and they can be recognised by how the observed brightness changes: short-lived blips against a constant background. They're useless as standard candles.

However, another type of variable star holds more promise: intrinsic variables. These are stars with nuclear reactions whose energy output fluctuates periodically, repeating the same pattern of changes over and over again. The light they emit also fluctuates. Intrinsic variables can also be recognised, because the changes in light output are *not* sudden blips.

Leavitt was studying a particular type of variable star, called a Cepheid because the first one discovered was Delta Cephei. Using an ingenious statistical approach, she discovered that brighter Cepheids have longer periods, according to a specific mathematical rule. Some Cepheids are near enough to have measurable parallax, so she could calculate how far away they are. From that she could calculate their intrinsic brightness. And those results then extended to all Cepheids, using the formula relating the period to the intrinsic brightness.

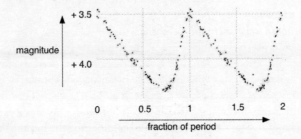

Observed light curve of Delta Cephei.

1 Pluto, depicted in false colour to amplify colour variations.

2 Pluto's satellite Charon.

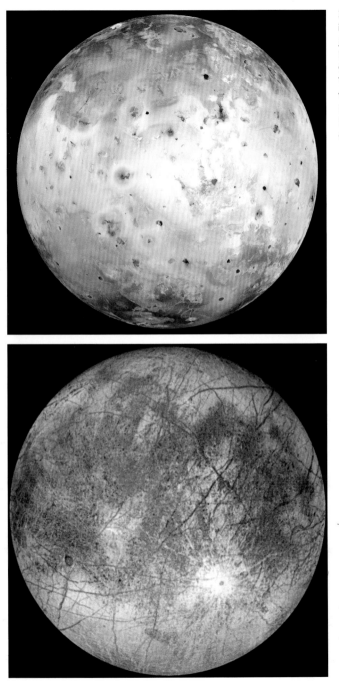

3 True colour image of Jupiter's moon Io taken by the Galileo probe. The erupting volcano Prometheus is just left of centre.

4 True colour image of Jupiter's moon Europa taken by the Galileo probe.

5 Saturn imaged by NASA's Cassini probe in 2006 looking back towards the Sun. The fuzzy outermost ring is the E ring, created by ice-fountains on the moon Enceladus which spew ice-particles into space.

6 Halley's comet.

7 Solar flare, 16 December 2014.

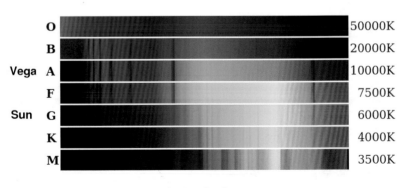

	O	50000K
	B	20000K
Vega	A	10000K
	F	7500K
Sun	G	6000K
	K	4000K
	M	3500K

8 A selection of stellar spectra.

9 The Milky Way in the direction of the galactic core.

10 The barred spiral galaxy NGC 1300.

11 Spiral galaxy NGC 1232.

12 Galactic cluster Abell 2218, 2 billion light years away in the constellation Draco. The gravitational fields of these galaxies act as lenses, distorting more distant galaxies into thin arcs.

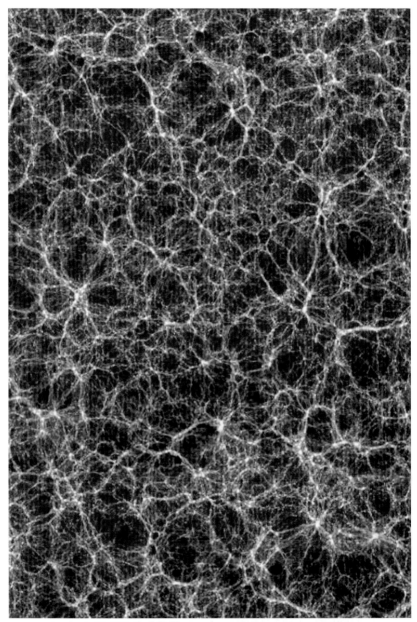

13 Computer simulation of the distribution of matter in the universe, showing skeins of matter separated by huge voids in a cube 2 billion light years across.

Cepheids were the long-sought standard candle. Together with the associated yardstick – the formula describing how a star's apparent brightness varies with distance – they allowed us to take another step up the cosmic distance ladder. Each step involved a mixture of observations, theory, and mathematical inference: numbers, geometry, statistics, optics, astrophysics. But the final step – a truly giant one – was yet to come.

Great Sky River

See yonder, lo, the Galaxy,
Which men call the Milky Way,
For it is white.

Geoffrey Chaucer, *The House of Fame*

IN ANCIENT TIMES there was no street lighting beyond the occasional torch or fire, and it was virtually impossible not to notice one amazing feature of the heavens. Today the only way to see it is to live in, or visit, a region where there's little or no artificial lighting. Most of the night sky is dusted with bright pinpoints of stars, but across the whole thing runs a broad, irregular band of light, more like a river than a scatter of glowing points. Indeed, to the ancient Egyptians it *was* a river, the celestial analogue of the Nile. Today we still call it the Milky Way, a name that reflects its puzzling form. Astronomers call the cosmic structure that creates it the Galaxy, a word derived from the ancient Greek names *galaxias* (milky one) and *kyklos galaktikos* (milky circle).

It took millennia for astronomers to realise that this milky smear across the sky is, despite appearances, a gigantic band of stars, so distant that the eye can't resolve them into individual points. This band is actually a lens-shaped disc, seen edge-on, and we're inside it.

As astronomers surveyed the heavens with ever more powerful telescopes, they noticed other faint smudges, quite unlike stars. A few are visible to a keen eye: the tenth-century Persian astronomer Abd al-Rahman al-Sufi described the Andromeda Galaxy as a small cloud, and in 964 he included the Large Magellanic Cloud in his *Book of*

Milky Way over Summit Lake, West Virginia.

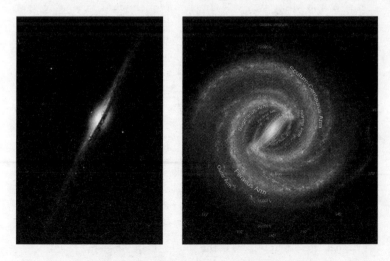

Left: A galaxy seen edge-on, with central bulge. Right:
Artist's impression of our own Milky Way Galaxy.

Fixed Stars. Originally, western astronomers called these faint, fuzzy wisps of light 'nebulas'.

We now call them galaxies. Ours is *the* Galaxy. They are the most numerous large structures that organise stars. Many display stunning patterns – spiral arms, whose origins remain controversial. Despite their ubiquity, galaxies have many features that we don't fully understand.

✦

In 1744 Charles Messier compiled the first systematic catalogue of nebulas. His first version contained 45 objects, and the 1781 version increased this to 103. Shortly after publication, Messier and his assistant Pierre Méchain discovered seven more. Messier noticed an especially prominent nebula in the constellation of Andromeda. It's known as M31 because it was the 31st nebula on his list.

Cataloguing is one thing, understanding quite another. What is a nebula?

As early as 400 BC the Greek philosopher Democritus suggested that the Milky Way is a band of tiny stars. He also developed the idea that matter is made from tiny indivisible atoms. Democritus's theory of the Milky Way was largely forgotten, until Thomas Wright published *An Original Theory or New Hypothesis of the Universe* in 1750. Wright revived the suggestion that the Milky Way is a disc of stars, too far away to be resolved as individuals. It also occurred to him that nebulas might be similar. In 1755 the philosopher Immanuel Kant renamed nebulas 'island universes', acknowledging that these cloudy smudges are composed of innumerable stars, even more distant than those in the Milky Way.

Between 1783 and 1802 William Herschel found another 2500 nebulas. In 1845 Lord Rosse used his new, impressively large telescope to detect individual points of light in a few nebulas, the first significant evidence that Wright and Kant might be right. But this proposal was surprisingly controversial. Whether these wisps of light are separate from the Milky Way – indeed, whether the Milky Way constitutes the entire universe – was still up for grabs as recently as 1920, when Harlow Shapley and Heber Curtis held the Great Debate in the Smithsonian Museum.

Shapley thought that the Milky Way is the whole universe. He argued that if M31 is like the Milky Way, it must be about 100 million light years away, a distance considered too big to be credible. In support, Adriaan van Maanen claimed he'd observed the Pinwheel Galaxy rotating. If it were as far away as Curtis's theory predicted, parts of it had to be moving faster than light. Yet another nail in the coffin was a nova observed in M31, a single exploding star that temporarily produced more light than the entire nebula. It was hard to see how one star could outshine a collection of millions of them.

Hubble settled the debate in 1924, thanks to Leavitt's standard candles. In 1924 he used the Hooker telescope, the most powerful available, to observe Cepheids in M31. Leavitt's distance–luminosity relation told him they were 1 million light years away. This placed M31 far beyond the Milky Way. Shapley and others tried to persuade him not to publish this absurd result, but Hubble went ahead: first in the *New York Times*, then in a research paper. It later turned out that van Maanen was wrong, and Shapley's nova was actually a supernova, which did produce more light than the galaxy it was in.

Further discoveries revealed that the Cepheid story is more complicated. Walter Baade distinguished two different kinds of Cepheid, each with a different period–luminosity relationship: classical Cepheids and Type II Cepheids, showing that M31 was even more distant than Hubble had stated. The current estimate is 2·5 million light years.

✦

Hubble was especially interested in galaxies, and he invented a classification scheme based on their visual appearance. He distinguished

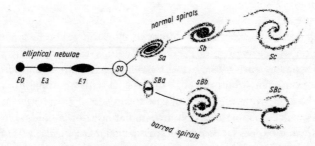

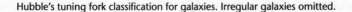
Hubble's tuning fork classification for galaxies. Irregular galaxies omitted.

four main types: elliptical galaxies, spiral galaxies, barred spirals, and irregular galaxies. Spiral galaxies in particular raise fascinating mathematical issues, because they show us the consequences of the law of gravity on a gigantic scale – and what emerges is an equally gigantic pattern. Stars seem to be scattered at random in the night sky, yet when you put enough of them together, you get a mysteriously regular shape.

Hubble didn't answer these mathematical questions, but he got the whole subject off the ground. One simple but influential contribution was to organise the shapes of galaxies into a diagram shaped like a tuning fork. He assigned symbolic types to these shapes: E1–E7 for the ellipticals, Sa, Sb, Sc for spirals, SBa, SBc, SBc, for barred spirals. His classification was empirical, that is, not based on any detailed theory or system of measurements. But many important branches of science have started as empirical classifications – among them geology and genetics. Once you have an organised list, you can start to work out how the different examples fit together.

For a time it was thought that perhaps the diagram illustrates the long-term evolution of galaxies, starting out as tight elliptical clusters of stars, then thinning out and developing either spirals or bars and spirals, depending on the combination of mass, diameter, and rotation speed. Then the spirals become ever more loosely wound, until the galaxy loses much of its structure and becomes irregular. It was an attractive idea, because the analogous Herzsprung–Russell diagram for the spectral types of stars does represent stellar evolution to some extent. However, it's now thought that Hubble's scheme is a catalogue of possible forms, and galaxies don't evolve in such a tidy manner.

✦

Compared to the featureless clumps of elliptical galaxies, the mathematical regularity of spirals and barred spirals stands out. Why are so many galaxies spiral? Where does the central bar in about half of them come from? Why don't the others have it as well? You might imagine that these questions are relatively easy to answer: set up a mathematical model, solve it, most probably by simulating it on a computer, and watch what happens. Since the constituent stars of a galaxy are fairly thinly spread, and don't move at near lightspeed, Newtonian gravity ought to be sufficiently accurate.

Many theories of that type have been studied. No definitive explanation has emerged as a front runner, but a few theories fit observations better than most. Only fifty years ago, most astronomers believed the spirals were caused by magnetic fields, but we now know these are too weak to explain spirals. Today there's general agreement that the spiral shape results mainly from gravitational forces. Exactly how is another matter.

One of the first theories to gain widespread acceptance was proposed by Bertil Lindblad in 1925, and it's based on a special type of resonance. Like Poincaré, he considered a particle in a nearly circular orbit in a rotating gravitational landscape. To a first approximation, the particle goes in and out relative to the circle with a specific natural frequency. A Lindblad resonance occurs when this frequency bears a fractional relationship to the frequency at which it encounters successive crests in the landscape.

Lindblad understood that a spiral galaxy's arms can't be permanent structures. In the prevalent model of how stars move in a galaxy, their speeds vary with radial distance. If the same stars stayed in the arm all the time, the arm would become ever more tightly wound, like overwinding a clock. Although we can't observe a galaxy for millions of years to see whether the winding gets tighter, there are lots of galaxies, and none of them seems overwound. He proposed that stars are repeatedly recycled through the arms.

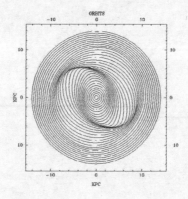

How stars orbiting in ellipses can create a spiral
density wave – here, a barred spiral.

In 1964 Chia-Chiao Lin and Frank Shu suggested that the arms are density waves, in which stars pile up temporarily. The wave moves on, engulfing new stars and leaving previous ones behind, much like an ocean wave travels across the sea for hundreds of miles, but carries no water with it (until it gets near land, when the water piles up and rushes ashore). The water just goes round and round as the wave passes. Lindblad and Per Olof Lindblad took up this idea and developed it further. It turns out that Lindblad resonances can create these density waves.

The main alternative theory is that the arms are shockwaves in the interstellar medium, where matter piles up, initiating star formation when it becomes sufficiently dense. A combination of both mechanisms is entirely feasible.

✦

These theories for the formation of spiral arms held sway for more than fifty years. However, recent mathematical advances suggest something very different. The key exhibit is barred spirals; these have the iconic spiral arms but there's a straight bar across the middle. A typical example is NGC 1365.

Barred spiral galaxy NGC 1365 in the Fornax galactic cluster.

One way to approach galaxy dynamics is to set up an n-body simulation with large values of n, modelling how each star moves in response to the gravitational pull of all the others. A realistic application of this method needs a few hundred billion bodies, and the computations aren't feasible, so simpler models are used instead. One of them provides an explanation for the regular pattern of the spiral arms. Paradoxically, they're caused by chaos.

If you think that 'chaos' is just a fancy word for 'randomness', it's hard to see how a regular pattern can have a chaotic explanation. This comes about because, as we saw, chaos isn't really random. It's a consequence of deterministic rules. In a sense, those rules are hidden patterns that underpin the chaos. In barred spiral galaxies, individual stars are chaotic, but as they move, the galaxy retains an overall spiral form. As stars move away from concentrations along the spiral arms, new ones take their place. The possibility of patterns in chaotic dynamics is a warning to scientists who assume that a patterned result must have a similarly patterned cause.

In the late 1970s George Contopoulos and coworkers modelled barred spiral galaxies by assuming a rigidly rotating central bar, and using n-body models to determine the dynamics of stars in the spiral arms, driven by the rotation of the central bar. This set-up builds in the bar morphology as an assumption, but it shows that the observed shape is sensible. In 1996 David Kaufmann and Contopoulos discovered that the inner parts of the spiral arms, apparently spinning off from the ends of the bar, are maintained by stars that follow chaotic orbits. The central region, especially the bar, rotates like a rigid body: this effect is called corotation. The stars that create the inner arms belong to the so-called 'hot population', wandering chaotically into and out of the central region. The outer arms are created by stars following more regular orbits.

The rotating bar has a gravitational landscape much like that for Poincaré's 2½-body problem, but the geometry is different. There are still five Lagrange points, where a speck of dust would remain at rest in a frame of reference that rotates along with the bar, but they're rearranged differently, in a cross shape. However, the model now includes around 150,000 dust specks – the other stars – and the specks exert forces on each other, though not on the bar. Mathematically, this is a 150,000-body simulation in a fixed but rotating gravitational landscape.

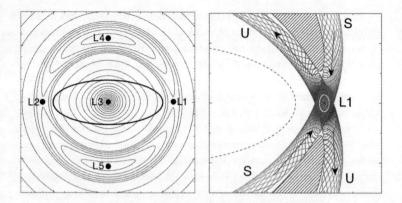

Left: Lagrange points for the rotating bar. Right: stable
(S) and unstable (U) manifolds near L1.

Three of the Lagrange points, L3, L4, and L5, are stable. The other two, L1 and L2, are unstable saddles, and they lie close to the ends of the bar, which is drawn as an ellipse. Now we need a quick dose of nonlinear dynamics. Associated with equilibria of saddle type are two special multidimensional surfaces, called stable and unstable manifolds. These names are traditional, but potentially confusing. They don't mean that the associated orbits are stable or unstable. They just indicate the direction of flow that defines these surfaces. A speck of dust placed on the stable manifold will approach the saddle point as if attracted to it; one placed on the unstable manifold will move away as if repelled. A particle placed anywhere else will follow a path combining both types of motion. It was by considering these surfaces that Poincaré made his initial discovery of chaos in the 2½-body problem. They intersect in a homoclinic tangle.

If all that mattered in the problem were position, the stable and unstable manifolds would be curves, crossing at the saddle point. The contour lines near L1 and L2 leave a cross-shaped gap, shown enlarged in the right-hand picture. These curves run through the middle of the gaps. However, astronomical orbits involve velocities as well as positions. Together these quantities determine a multidimensional space called *phase space*. Here the two dimensions of position, shown directly in the picture, must be complemented by two more dimensions of velocity. The phase space is *four*-dimensional, and the stable and unstable manifolds are two-dimensional surfaces, illustrated in the right-hand picture as tubes marked with arrows. S is the stable manifold, U the unstable one.

Where these tubes meet, they act like gateways between the corotation region and its exterior. Stars can pass in and out along them, in the directions shown by the arrows, and they can swap tubes chaotically where they cross. So some stars from inside the corotation region pass through this gate and stream off along the tube marked U, bottom right. Now a phenomenon known as 'stickiness' comes into play. Although the dynamics is chaotic, stars that exit through this gateway hang around close to the unstable manifold for a long time – possibly longer than the age of the universe. Putting it all together, stars stream out near L1 and then follow the outward-bound branch of the unstable manifold, which here turns clockwise. The same happens at L2, 180 degrees round the galaxy, and again the stream goes clockwise.

Eventually many of these stars are recycled back into the corotation region, and everything happens all over again, though not at regular intervals of time because of the chaos. So what we see is a pair of spiral arms that emerge from the ends of the bar, at an angle, while the whole shape rotates steadily. Individual stars don't remain at fixed places in the arms. They're more like the sparks thrown off by a Catherine wheel (pinwheel) firework as it spins. Except these sparks eventually come back to the middle to be spat out again, and their paths vary chaotically.

The left-hand picture shows the positions of the stars at one typical time in an *n*-body simulation of this model. Two spiral arms and the central bar are evident. The right-hand picture shows the corresponding unstable manifolds, which match the densest regions in the left-hand picture. The next picture shows which parts of the galaxy are occupied by stars in various populations of regular and chaotic orbits. Regular orbits are confined to the corotation zone, as we'd expect; chaotic ones occur there too, and they dominate outside it, where the spiral arms are.

It's worth comparing this theory with the twisted series of ellipses in the picture on page 177. The ellipses put in a pattern in order to get one out. However, real *n*-body dynamics doesn't yield elliptical orbits, because all the bodies perturb each other, so the proposed pattern doesn't really make sense, unless it's a reasonable approximation to

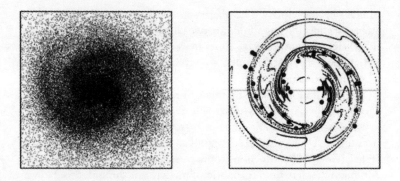

Left: Projection of the n-body system on the galactic plane. Right: spiral pattern with the unstable invariant manifolds emanating from L1 and L2.

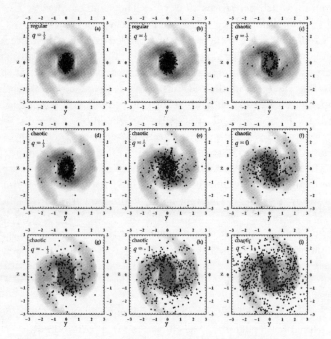

Instantaneous positions of particles belonging to different populations of the regular and chaotic orbits (black dots) superimposed on the backbone of the galaxy on the plane of rotation (grey background).

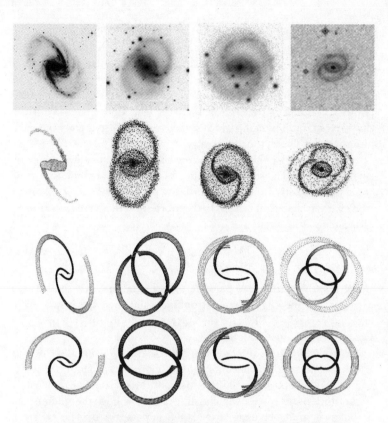

Ring and spiral-arm morphologies. Top row: Four galaxies, namely NGC 1365, NGC 2665, ESO 325–28, and ESO 507–16. Second row: Schematic plots of these galaxies, bringing out the spiral and ring structures. Third row: Examples from stable/unstable manifold calculations with similar morphology, projected in roughly the same way as the observed galaxy or schematic plot. Fourth row: Face-on view of these manifolds with the bar along the x-axis.

something that does. The chaotic model does build the central bar in as an explicit assumption, but everything else emerges from genuine *n*-body dynamics. What we get is chaos – as we might expect – but also the spiral pattern, created by the chaos. There's a message here: take the mathematics seriously, and the patterns will take care of themselves. Force patterns artificially, and you risk getting nonsense.

There's further confirmation that sticky chaos plays a role in the formation of spirals in barred spiral galaxies. This is the common presence, in such galaxies, of rings of stars, very regular in form, often overlapping in pairs. Again, the idea is that in such galaxies, sticky chaos aligns many stars with the unstable manifolds of the Lagrange points L_1 and L_2 at the ends of the bars. This time we also consider the stable manifolds, along which stars return to the gateways and back into the core. These, too, are sticky.

The top row of the next picture shows four typical examples of ringed galaxies. The second row shows drawings that emphasise their spiral and ring structures. The third row provides matching examples from the mathematical model. In the fourth row, these are seen face-on rather than at an angle.

✦

Using a spectroscope, it's possible to estimate how fast the stars in a galaxy are moving. When astronomers did this, the results were extremely puzzling. I'll leave the current resolution of this puzzle to Chapter 14, and just set it up here.

Astronomers measure how fast galaxies rotate using the Doppler effect. If a moving source emits light of a given wavelength, the wavelength changes according to the velocity of the source. The same effect occurs for sound waves; the classic example is how the pitch of an ambulance siren becomes lower when it passes you. The physicist Christian Doppler analysed this effect in 1842 in a paper about binary stars, using Newtonian physics. A relativistic version predicts the same basic behaviour, but with quantitative differences. Light comes in many wavelengths, of course, but spectroscopy shows up particular wavelengths as dark lines in the spectrum. When the source is moving, those lines shift by a uniform amount, and it's straightforward to calculate the velocity from the amount of shift.

For galaxies, the standard spectral line used for this purpose is the hydrogen-alpha line. For a stationary source it sits in the deep red part of the visible spectrum, and it arises when an electron in an atom of hydrogen switches from its third lowest energy level to the second lowest. Hydrogen is the commonest element in the universe, so the hydrogen-alpha line is often prominent.

It's even possible – for galaxies not *too* far away – to measure the rotation speed at different distances from the centre of the galaxy. These measurements determine the galaxy's rotation curve, and it turns out that the speed of rotation depends only on the distance from the centre. To a good approximation, a galaxy moves like a series of concentric rings, each spinning rigidly but at a speed that can vary from ring to ring. This is reminiscent of Laplace's model for Saturn's rings (Chapter 6).

In this model, Newton's laws lead to a key mathematical exhibit: a formula relating the rotation speed at a given radius to the total mass inside that radius. (Stars move so slowly compared to lightspeed that relativistic corrections are generally thought to be irrelevant.) It states that the total mass of a galaxy, out to a given radius, is that radius, multiplied by the square of the rotational velocity of the stars at that distance, and divided by the gravitational constant.[1] This formula can be rearranged to express the rotational velocity at a given radius: multiply the total mass inside that radius by the gravitational constant, divide by the radius, and take the square root. This formula, in either version, is called the Kepler equation for the rotation curve, because it can also be derived directly from Kepler's laws.

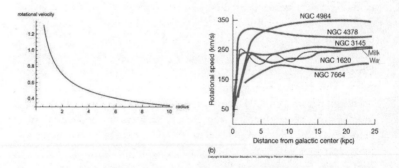

Left: Rotation curve predicted by Newtonian laws. Scales in arbitrary units. Right: Observed rotation curves for six galaxies.

It's hard to measure the mass distribution directly, but one prediction is independent of such considerations: how the rotation curve behaves for large enough radii. Once the radius approaches the observed radius of the galaxy, the total mass inside that radius becomes almost constant, equal to the total mass of the galaxy. So when the radius is large enough, the rotational speed is proportional to the reciprocal of the square root of the radius. The left-hand picture is the graph of this formula, which decays towards zero as the radius grows.

For comparison, the right-hand picture shows observed rotation curves for six galaxies, one of them ours. Instead of decaying, the rotation speed grows, and then stays roughly constant.

Oops.

Alien Worlds

Alien astronomers could have scrutinized Earth for more than 4 billion years without detecting any radio signals, despite the fact that our world is the poster child for habitability.

Seth Shostak, *Klingon Worlds*

IT'S LONG BEEN AN ARTICLE OF FAITH among science fiction writers that the universe is littered with planets. This belief was motivated mainly by narrative imperative: planets are needed as locations for exciting stories. However, it always made good scientific sense. Given the amount of cosmic junk that rattles around inside the universe, of all shapes and sizes, there should be planets in abundance.

As far back as the sixteenth century, Giordano Bruno stated that stars are distant suns with planets of their own, which might even be inhabited. A thorn in the side of the Catholic Church, he was burned at the stake for heresy. At the end of his *Principia*, Newton wrote: 'If the fixed stars are the centres of similar systems [to the solar system], they will all be constructed according to a similar design...'

Other scientists disagreed, maintaining that the Sun is the only star in the universe to possess planets. But the smart money was always on zillions of exoplanets, as they're called. Our best theory of planet formation is the collapse of a vast gas cloud, which forms planets at the same time as their central star, and such clouds exist in abundance. There are at least 50 quintillion large bodies – stars – and a far greater number of small ones – dust particles. It would be strange if there were some forbidden intermediate size range, and even stranger if it happened to coincide with the typical sizes of planets.

✦

Indirect arguments are all very well, but the elephant in the room was notable by its absence. Until very recently, there was no observational evidence that any other star has planets. In 1952 Otto Struve suggested a practical method for detecting exoplanets, but forty years were to pass before it came to fruition. In Chapter 1 we saw that the Earth and the Moon behave like a fat man dancing with a child. The child spins round and round while the man pivots on his feet. The same goes for a planet orbiting a star: the lightweight planet moves in a big ellipse while the hefty star wobbles slightly.

Struve suggested using a spectroscope to detect these wobbles. The Doppler effect makes any movement of the star shift its spectral lines slightly. The amount of shift gives the velocity of the star; we deduce the presence of the whirling child by watching how the fat man wobbles. This method works even if there are several planets: the star still wobbles, but in a more complicated way. The picture shows how the Sun wobbles. Most of the movement is caused by Jupiter, but other planets contribute too. The overall movement is about three times the Sun's radius.

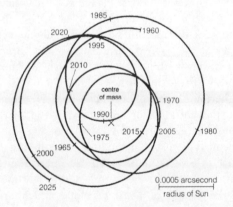

Motion of the Sun relative to the centre of
mass of the solar system, 1960–2025.

Struve's technique of Doppler spectroscopy led to the first confirmed exoplanet sighting in 1992, by Aleksander Wolszczan and Dale Frail. The primary is a curious type of stellar object known as a pulsar. These bodies emit rapid regular radio pulses. We now explain them as rapidly spinning neutron stars, so named because most of their matter is neutrons. Wolszczan and Frail used radio astronomy to analyse tiny variations in the pulses emitted by pulsar PSR 1257+12, and deduced that it has at least two planets. These alter its spin slightly and affect the timing of the pulses. Their result was confirmed in 1994, along with the presence of a third planet.

Pulsars are rather unusual, and reveal nothing significant about ordinary stars. But those, too, started to yield up their secrets. In 1995 Michel Mayor and Didier Queloz discovered an exoplanet orbiting 51 Pegasi, a star in the same general spectral class (G) as the Sun. Later, it turned out that both groups had been pipped to the post in 1988, when Bruce Campbell, Gordon Walker, and Stephenson Yang noticed that the star Gamma Cephei was wobbling suspiciously. Because their results were at the limit of what could be detected, they hadn't claimed sighting a planet, but within a few years more evidence came in and astronomers started to believe the group had done just that. Its existence was finally confirmed in 2003.

We now know of more than two thousand exoplanets – the current figure (1 June 2016) is 3422 planets in 2560 planetary systems, including 582 systems with more than one planet. Plus thousands of plausible candidates yet to be confirmed. However, occasionally what was thought to be a sign of an exoplanet is re-examined and discarded as something else, and new candidates are flooding in, so these figures can go down as well as up. In 2012 it was announced that one member of the nearest star system, Alpha Centauri, has a planet – Earth-sized but much hotter.[1] It now looks as though this planet, Alpha Centauri Bb, doesn't actually exist, but is an artefact of the data analysis.[2] However, another potential exoplanet Alpha Centauri Bc round the same star has since been detected. Gliese 1132, a red dwarf star 39 light years away, definitely has a planet, GJ 1132b, which has caused much excitement because it's Earth-sized (though far too hot to have liquid water) and close enough to observe its atmosphere.[3] Many exoplanets lie within a few dozen light years. Planetarily, we're not alone.

Initially the only worlds that could be observed were 'hot Jupiters':

massive planets very close to their stars. This tended to give a biased impression of the type of world that's out there. But the techniques are becoming more sensitive at a rapid pace, and we can now detect planets the size of the Earth. We can also begin to find out whether they have atmospheres or water, using spectroscopy. Statistical evidence indicates that planetary systems are commonplace throughout the Galaxy – indeed, the entire universe – and Earthlike[4] planets in Earthlike orbits around Sunlike stars, while a small proportion, exist in their billions.

✦

There are at least a dozen other ways to detect exoplanets. One is direct imaging: aiming a very powerful telescope at a star and spotting a planet. It's a bit like trying to see a match in the glare of a searchlight, but clever masking techniques that remove the star's own light sometimes make it possible. The commonest way to detect exoplanets is the transit method. If it so happens that a planet crosses the disc of the star, as viewed from Earth, then it blocks out a small part of the star's light output. The transit creates a characteristic dip in the light curve. Most exoplanets are unlikely to be oriented so favourably, but the proportion of those that create transits is big enough for the approach to be viable.

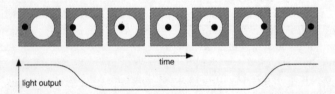

Simple model of how the star's light output becomes dimmer when a planet transits. Assuming the star emits the same amount of light at each point, and that the planet blocks all of it, the light curve remains flat while the whole of the planet blocks the light. In practice these assumptions are not quite correct, and more realistic models are used.

The picture is a simplified illustration of the transit method. As the planet begins its transit, it starts to block some of the light from the star. Once the entire disc of the planet lies within that of the star, the light output levels off, and remains roughly constant until the planet approaches the other edge of the star. As the planet exits across the edge, the star returns to its original apparent brightness. In practice the star generally appears less bright near its edges, and some light may be deflected round the planet if it has an atmosphere. More sophisticated models correct for these effects. The picture shows a real light curve (dots) for a transit of exoplanet XO-1b across the star XO-1, together with a fitted model (solid curve).

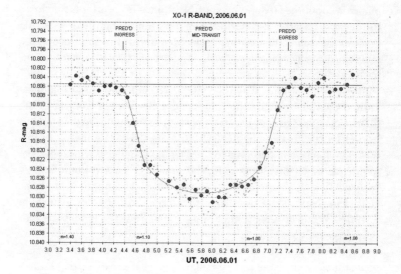

Light curve for the 1 June 2006 exoplanet transit of the 10·8 R-magnitude star XO-1 by the Jupiter-sized planet XO-1b. Solid dots are 5-point averages of magnitudes from images shown by small dots. The line is a fitted model.

The transit method, analysed mathematically, provides information on the size, mass, and orbital period of the planet. It sometimes tells us about the chemical composition of the planet's atmosphere, by comparing the star's spectrum with light reflected from the planet.

✦

NASA chose the transit method for its Kepler telescope – a photometer that measures light levels with exquisite accuracy. Launched in 2009, Kepler kept an eye on the light output of more than 145,000 stars. The plan was to observe them for at least three and a half years, but Kepler's reaction wheels, used to keep the telescope aligned, began to fail. In 2013 the mission design was changed so that the disabled instrument could still do useful science.

The first exoplanet that Kepler found, in 2010, is now called Kepler-4b. Its parent star is Kepler-4, about 1800 light years away in the constellation Draco, a star similar to the Sun but a bit bigger. The planet is about the size and mass of Neptune, but its orbit is far closer to the star. Its orbit has period 3·21 days, and a radius of 0·05 AU – about one tenth the distance from Mercury to the Sun. Its surface temperature is a sweltering 1700 K. The orbit is eccentric, with eccentricity about 0·25.

Despite its faltering reaction wheels, Kepler found 1013 exoplanets around 440 stars, plus another 3199 candidates yet to be confirmed. Big planets are easier to spot because they block more light, so they tend to be over-represented among Kepler's exoplanets, but to some extent this tendency can be compensated for. Kepler has found enough exoplanets to provide statistical estimates for the number of planets in the Galaxy with particular characteristics. In 2013 NASA announced that the Galaxy probably contains at least 40 billion Earth-sized exoplanets in Earthlike orbits around Sunlike stars and red dwarfs. If so, the Earth is far from unique.

The catalogue of orbits and stellar systems contains many that seem totally unlike the solar system. Tidy patterns like the Titius–Bode law seldom hold. Astronomers are just beginning to grapple with the complexities of comparative anatomy of stellar systems. In 2008 Edward Thommes, Soko Matsumura, and Frederic Rasio simulated accretion from protoplanetary discs.[5] The results suggest that systems like our own are comparatively rare, occurring only when the variables that characterise the main features of the disc have values perilously close to those for which planets don't form at all. Giant planets are more common. In the parameter space of protoplanetary discs, ours skated on the edge of disaster. Some basic mathematical principles still apply, however: in particular, the occurrence of orbital resonances. For example, the star systems Kepler-25, Kepler-27, Kepler-30, Kepler-31 and Kepler-33 all have at least two planets in 2:1 resonance. Kepler-23, Kepler-24, Kepler-28 and Kepler-32 have at least two planets in 3:2 resonance.

✦

The planet hunters are already adapting their techniques to search for other features of star systems, among them exomoons and exoasteroids, which can add tiny extra blips to light curves in a very complicated way. David Kipping is using a supercomputer to re-examine Kepler data on 57 exoplanet systems, seeking hints of an exomoon. René Heller has performed theoretical calculations suggesting that an exoplanet several times larger than Jupiter (not uncommon) could have a Mars-sized moon, and in principle Kepler could spot this. Jupiter's moon Io causes radio bursts by interacting with the planet's magnetic field, and similar effects might happen elsewhere, so Joaquin Noyola is searching for exolunar radio signals. When NASA's James Webb space telescope, successor to the Hubble, launches in 2018 it might be able to image an exomoon directly.

Michael Hippke and Daniel Angerhausen have been hunting exotrojans. Recall that a Trojan asteroid follows a planet in much the same orbit, spaced 60 degrees ahead or 60 degrees behind, so it creates its own tiny blip as it crosses the star. Astronomers have looked for these blips, but nothing has yet been spotted because the effects would be very small. Instead, Hippke and Angerhausen use a statistical approach, like

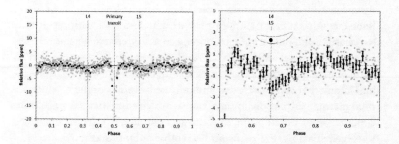

Left: Combined light curves for a million transits, showing small dips at the Trojan points L4 and L5 (marked). These are not statistically significant. Right: 'Folded' data show a statistically significant dip.

wandering through a game reserve counting lion tracks. They don't tell you which lions made them, but you can estimate how common lions are. They combined nearly a million light curves to enhance the signals associated with exotrojans. The results show slight blips at the Trojan points, but these aren't statistically significant. However, if the plot is folded in half, so that the positions a given number of degrees ahead of the orbit and behind it coincide, there's a statistically significant blip at 60 degrees (plus or minus combined).[6]

✦

Science fiction's widespread assumption that distant stars often have planets, though once derided, has turned out to be absolutely correct. What about a related science fictional trope: the existence of intelligent[7] alien life forms? This is a much more difficult issue, but again it would be bizarre if a universe with quintillions of planets managed to produce precisely *one* with intelligent life. Too many factors have to balance precisely to make our world unique.

In 1959, in the journal *Nature*, Giuseppe Cocconi and Philip Morrison published a provocative article, 'Searching for interstellar communications'. They pointed out that radio telescopes had become sensitive enough to pick up a radio message from an alien civilisation. They also suggested aliens would choose a landmark frequency: the 1420 Mhz HI line in the hydrogen spectrum. This is special because hydrogen is the commonest element in the universe.

The radio astronomer Frank Drake decided to test Cocconi and Morrison's idea by initiating Project Ozma, which searched for such signals from the nearby stars Epsilon Eridani and Tau Ceti. He didn't detect anything, but in 1961 he organised a conference on the 'search for extraterrestrial intelligence'. At that meeting he wrote down a mathematical equation expressing the number of alien civilisations in our Galaxy that can currently communicate by radio, as the product of seven factors, such as the average rate at which stars form, the fraction of planets that develop life, and the average time over which civilisations have the ability to transmit detectable radio signals.

The Drake equation is often used to calculate how many alien civilisations, able to communicate, exist, but that wasn't Drake's intention. He was trying to isolate the important factors that scientists should focus on. His equation has flaws, if you take it literally, but thinking about them provides insight into the likelihood of alien civilisations and the possibility that we could detect their signals. A major successor to Project Ozma is SETI, the Search for Extraterrestrial Intelligence, founded in 1984 by Thomas Pierson and Jill Tarter to spearhead a systematic search for alien communications.

The Drake equation isn't terribly practical, because it's highly sensitive to errors. 'Planet' may be too restrictive, as we'll see shortly. So may 'radio'. Expecting aliens to communicate by outmoded radio technology could be as pointless as watching for smoke signals. Even more questionable is the idea that we would recognise their communications as communications. With the advent of digital electronics, most of our own signals, even over mobile phones, are digitally encoded to compress their information and eliminate errors caused by extraneous noise. Aliens would surely do likewise. In 2000 Michael Lachmann, Mark Newman, and Cris Moore proved that efficiently coded communications look exactly the same as random black-body radiation. This is the spectrum of electromagnetic radiation from an opaque, non-reflective body at a constant temperature. Their paper originally bore the title 'Any sufficiently advanced communication is indistinguishable from noise'.[8]

✦

Intelligence is aiming high. Even unintelligent alien life would be a game changer.

When assessing the chances of alien life, it's all too easy to fall into the trap of imagining that the perfect place for aliens must be an Earthlike planet – one that's about the same size as ours, at a similar distance from a similar star, whose surface is a mixture of rocky land and liquid water (like ours), and whose atmosphere contains oxygen (ditto). Yes, Earth is the only inhabited planet we know about, but we've only just started to look. Within the solar system, all other worlds seem barren and inhospitable – though, as we'll see, that judgement shouldn't be made too hastily. So the best place to look for life seems to be outside the solar system.

The chances of life existing elsewhere are improved by a basic biological principle: life *adapts* to the prevailing conditions. Even on Earth, living creatures occupy an astonishing variety of habitats: deep in the oceans, high in the atmosphere, in swamps, in deserts, in boiling springs, beneath the Antarctic ice, and even three kilometres underground. It seems reasonable that alien life forms might occupy an even wider range of habitats. *We* might not be able to live there, but humans can't actually survive unaided in most terrestrial habitats. What is habitable depends on what's inhabiting it.

Our very terminology betrays deep prejudices. In recent years biologists have discovered bacteria that can live in boiling water, and other bacteria that survive in very cold conditions. Collectively, they're called extremophiles – creatures that like extremes. They're often portrayed as if they're clinging on precariously in a hostile environment, liable to go extinct at any moment. In reality, however, these creatures are beautifully adapted to their environment, and would die if transported into ours. Compared with them, *we* are the extremophiles.[9]

All life on Earth is related; it seems to have evolved from a single primordial biochemical system. As such, 'life on Earth' in all its rich variety reduces to a *single* data point. The Copernican principle maintains that there's nothing terribly special about human beings or their surroundings. If so, our planet is unlikely to be special – but that doesn't imply that it must be typical. Biochemists have made unusual variants of the molecules that form the basis of terrestrial genetics: DNA, RNA, amino acids, proteins, to find out whether the molecules used on Earth are the only ones that work. (They're not.) Such questions often lead to mathematical modelling as well as biology, because we can't be sure that biology elsewhere will be the same as it is here. It might use different chemistry, radically different chemistry, or avoid chemistry altogether by not being molecular.

That said, it makes excellent sense to *start* from that one genuine data point, as long as we don't forget that this is just a first step towards more exotic possibilities. That leads inevitably to one of the immediate goals of the planet hunters: to find an Earthlike exoplanet.

In astrobiological circles, a great deal of fuss is made about the so-called 'habitable zone' around a star. The habitable zone is not the region that might be habitable. It's the region around a star within which a hypothetical planet with enough atmospheric pressure could support liquid water. Get too close to the star and the water evaporates as steam; too far away and it freezes into ice. In between, the temperature is 'just right', and inevitably this region has acquired the nickname 'Goldilocks zone'.

The habitable zone of the solar system lies between about 0·73 and 3 AU from the Sun – the precise numbers are debatable. Venus just grazes the inner edge, the outer edge extends as far as Ceres, while Earth and Mars are safely inside. So 'in principle' the surfaces of Venus and Mars could support liquid water. In practice, though, it's more complicated. The mean surface temperature on Venus is 462°C, hot enough to melt lead, because Venus experienced a runaway greenhouse effect, trapping heat in its atmosphere. Liquid water seems highly implausible, to say the least. On Mars, the temperature is *minus* 63°C, so Mars was generally thought to have only solid ice. However, in 2015 it was discovered that small amounts of the ice melt in the Martian summer, seeping down the sides of some craters. This had been suspected for some time, because dark streaks are visible, but the crucial evidence is the presence of hydrated salts in the summer, when the streaks are growing longer. Mars probably had plenty of surface water about 3·8 billion years ago, but then it lost much of its atmosphere, blown away by the solar wind when the planet's magnetic field weakened. Some water evaporated and the rest froze. Mostly, it stays that way.

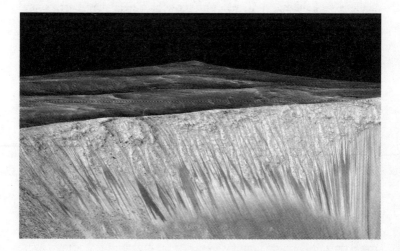

Dark streaks in Garni crater, on Mars, are caused by liquid water.

Distance from the primary, then, isn't the only criterion. The concept of a habitable zone provides a simple, comprehensible guideline, but guidelines aren't rigid. Liquid water need not exist inside the habitable zone, and can exist outside it. A planet that's close to its star may be in a zone that's far too hot, but if it's in 1:1 spin–orbit resonance then one side always faces the star, so it's hot, while the far side is extremely cold. In between is a temperate zone at right angles to the equator. (There's frozen ice on scalding hot Mercury, hidden away in polar craters where sunlight never penetrates. And that's not even in 1:1 resonance.) A planet with an icy surface might have some internal heat source – after all, the Earth does – that melts some of the ice. A thick atmosphere with lots of carbon dioxide or methane would also warm it up. A wobbly axis can help a planet stay warm outside the habitable zone by distributing heat unevenly. A planet with an eccentric orbit can store energy when it's near its star, releasing it as it moves away, even though on average it's not in the habitable zone. A red dwarf star can have a nearby planet with a thick cloudy atmosphere, redistributing heat more evenly.

In 2013 the Kepler telescope discovered two exoplanets that were the most similar to Earth then found. They both orbit the same star, Kepler-62 in the constellation Lyra, and their names are Kepler-62e and Kepler-62f. Each has a diameter about 50% bigger than Earth's, and they may be examples of super-Earths – rocky bodies more massive than the Earth but not as massive as Neptune. Alternatively, they could be compressed ice. They are firmly in Kepler-62's Goldilocks zone, so given suitable surface conditions such as an atmosphere similar to ours, they could have liquid water.

Early in 2015 NASA announced the discovery of two new exoplanets that resemble Earth even more closely. Kepler 438b is 12% larger than the Earth and receives 40% more energy from its star, which is 479 light years away. Kepler 442b is 30% larger than the Earth and receives 30% less energy; its star is 1292 light years away. It's not possible to confirm their existence by detecting corresponding wobbles in their stars. Instead, astronomers use careful comparisons of measurements, and statistical inference. Based on their size they're probably rocky, although their masses aren't known. Orbiting inside the habitable zone, they might have liquid water.

Other confirmed exoplanets that resemble Earth include Gliese

667Cc and 832c, and Kepler 62e, 452b, and 283c. An as yet unconfirmed Kepler candidate KOI-3010.01 is also Earthlike, if it exists. There are plenty of worlds resembling ours – not far away by cosmic standards, but inaccessible with current or foreseeable technology.

Peter Behroozi and Molly Peeples have reinterpreted the statistics of Kepler exoplanets in the context of our knowledge of how stars arise in galaxies, deriving a formula for how the number of planets in the universe changes as time passes.[10] The proportion of Earthlike worlds can be derived from this figure. Plugging in the current age of the universe, they estimate that there are roughly 100 quadrillion Earthlike planets at the present time. That's about 500 million per galaxy, so our Galaxy probably hosts half a billion planets very like our own.

The astrobiological focus is currently shifting from literally Earthlike planets to other types of world that might reasonably support life. According to simulations carried out by Dimitar Sasselov, Diana Valencia, and Richard O'Connell, super-Earths may be *more* suitable for life than our own planet.[11] The reason is plate tectonics. The movement of Earth's continents helps keep the climate stable by recycling carbon dioxide via the ocean floor, subduction, and volcanoes. Liquid water is more likely to hang around if the climate is stable, and that gives water-based life more time to evolve. So continental drift can improve the habitability of a planet.

Sasselov's team discovered that, contrary to expectations, continental drift is probably common elsewhere, and can happen on planets larger than Earth. The plates would be thinner than they are here, and move faster. So a super-Earth should have a more stable climate than we do, making it easier for complex life to evolve. The likely number of planets like Earth is quite big, but comparatively speaking such worlds are rare. However, there ought to be far more super-Earths, which greatly improves the prospects for *Earthlike* life. So much for 'rare Earth'. Further, Earth is not 'just right' for plate tectonics. We just scraped into the lower end of the suitable size range.

So much for Goldilocks.

✦

Maybe life doesn't need a planet at all.

Let's not give up on our own star system too easily. If life exists elsewhere in the solar system, where is it most likely to be? As far as we know, the only inhabited planet in the Sun's habitable zone is the Earth, so at first sight the answer has to be 'nowhere'. Actually, the places most likely to support life – probably no more complex than bacteria, but life nonetheless – are Europa, Ganymede, Callisto, Titan, and Enceladus. Ceres and Jupiter are outside bets.

Ceres, a dwarf planet, is on the outer edge of the habitable zone, and it has a thin atmosphere with water vapour. The *Dawn* mission revealed bright spots inside a crater, initially thought to be ice, but now known to be a type of magnesium salt. If it had been ice, Ceres would have had one key ingredient for Earthlike life, albeit frozen. Ice probably exists at greater depths.

Carl Sagan suggested in the 1960s that bacterial life, and possibly more complex balloon-like organisms, might float in Jupiter's atmosphere. The main obstacle is that Jupiter emits a lot of radiation. However, some bacteria thrive high in the Earth's atmosphere, where radiation levels are high, and tardigrades – small creatures popularly known as water bears – can survive levels of radiation and temperature extremes, both hot and cold, that would kill us.

The other five bodies I listed aren't planets or dwarf planets, but moons, and they lie well outside the habitable zone. Europa, Ganymede, and Callisto are moons of Jupiter. As discussed in Chapter 7, they have underground oceans, created because tidal heating by Jupiter melts the ice. There may well be hot hydrothermal vents at the bottom of these oceans, providing a habitat for life not unlike similar vents on Earth, such as those found along the Atlantic ridge. Here the Earth's tectonic plates are splitting apart, dragged asunder by a geological conveyor belt as their outer edges are subducted beneath the continents of Europe and America. The rich brew of volcanic chemicals, together with the warmth from hot volcanic gases, provides a comfortable habitat for tube worms, shrimps, and other quite complex organisms. Some evolutionary biologists think that life on Earth originated near such vents. If that works here, why not on Europa?

✦

Next in line is the most mysterious moon of them all, Saturn's satellite Titan. Its diameter is half as big again as the Moon's, and unlike any other moon in the solar system it has a dense atmosphere. The main body of Titan is a mixture of rock and water ice, and the surface temperature is about 95 K (minus 180°C). The *Cassini* mission revealed that it has lakes and rivers of liquid methane and ethane, which are gases at room temperature on Earth. The bulk of the atmosphere (98·4%) is nitrogen, plus 1·2% methane, 0·2% hydrogen, and traces of other gases such as ethane, acetylene, propane, hydrogen cyanide, carbon dioxide, carbon monoxide, argon, and helium.

Many of these molecules are organic – carbon based – and some are hydrocarbons. They're thought to be produced when ultraviolet light from the Sun breaks up methane, creating a dense orange smog. That's a puzzle in its own right, because it should take a mere 50 million years for all the methane in the atmosphere to break up – yet it's still there. Something must be replenishing it. Either volcanic activity is releasing methane from some vast underground reservoir, or the excess methane is produced by some exotic, probably primitive, organism. Out-of-balance chemistry is one potential sign of life; an obvious example is Earth's oxygen, which would have disappeared long ago were it not for photosynthesis by plants.

If Titan does have life, it must be radically different from life on Earth. The real point of the nursery tale is not that Goldilocks's preference was 'just right', but that Mummy Bear and Daddy Bear wanted what suited *them*, and it was different. It's the Bears' viewpoint that poses the most interesting and important scientific questions. Titan has no liquid water, though it does have ice pebbles. It's often assumed that water is essential for life, but astrobiologists have established that in principle lifelike systems could exist without water.[12] Titanian organisms could use some other fluid to convey important molecules around their bodies. Liquid ethane or liquid methane are possibilities: both can dissolve many other chemicals. A hypothetical Titanian could get its energy from hydrogen by reacting it with acetylene, giving off methane.

This is a typical example of 'xenochemistry' – possible chemical pathways in alien life forms, very different from the Earth norm. It shows that plausible organisms need not be similar to those on this planet, opening up more imaginative possibilities for alien life. However, chemistry alone won't create life. You need organised chemistry, very probably carried out in something along the lines of a cell. Our cells are surrounded by a membrane formed from phospholipids

– compounds of carbon, hydrogen, oxygen, and phosphorus. In 2015 James Stevenson, Jonathan Lunine, and Paulette Clancy put forward an analogue of a cell membrane that works in liquid methane, made from carbon, hydrogen, and nitrogen instead.[13]

✦

If humans evolved on Mars, how would they differ from us?

Silly question. Humans *didn't* evolve on Mars. If life evolved on Mars (and for all we know it might have done, long ago, and organisms on the level of bacteria may still exist there) it would follow its own evolutionary pathway, a mixture of accident and selective dynamics. If we transplanted humans to Mars, they'd die before they could evolve to suit the conditions there.

Very well, then. Suppose aliens evolve on some exoplanet. What would they look like? This is marginally more sensible. Bear in mind that Earth currently hosts millions of different species. What do *they* look like? Some have wings, some have legs, some have both, some live kilometres down in the oceans, some thrive in frozen wastes, others in deserts... Even Earthlike life is very diverse with weird biology – yeast has twenty sexes, *Xenopus* frogs eat their own young...

Movie and television aliens tend to be humanoid, allowing actors to portray them, or computer-generated monsters, designed for horrific effect. Neither is a reliable guide to likely alien life. Life evolves to suit the prevailing conditions and environment, and it's very diverse. We can speculate, of course, but no specific 'design' of an alien creature is likely to arise anywhere in the universe. The reason is a basic distinction in xenoscience, emphasised long ago by Jack Cohen: that between a universal and a parochial.[14] Here both are nouns, not adjectives, and are short for 'universal/parochial feature'. A parochial is a special feature that evolves by accident of history. For example, the human foodway crosses its airway, causing a number of deaths every year from the inhalation of peanuts. The number of fatalities is too small for this design flaw to have evolved away; it goes back to ancient fishy ancestors in the sea, where it didn't matter.

In contrast, a universal is a generic feature that offers evident survival advantages. Examples include the ability to detect sound or light, and the ability to fly in an atmosphere. One sign of a universal is that it has evolved several times independently on Earth. For example, flight has evolved in insects, birds, and bats, by independent routes. These routes differ in their parochial aspects; they all use wings, but each design of wing is very different. But they've all been selected for the same underlying universal.

This test has a defect, however: it links the feature directly to Earth's evolutionary history. Not so good when thinking about aliens. For example, is human-level (or higher) intelligence a universal? Intelligence has evolved independently in dolphins and octopuses, for instance, but not to our level, so it's not clear whether intelligence satisfies the 'multiply evolved' test. However, intelligence certainly looks like a generic trick that *could* evolve independently, and it offers clear short-term survival advantages by giving its possessor power over their environment. So intelligence is arguably a universal.

These aren't definitions, and the distinction between universals and parochials is fuzzy at best. But it focuses attention on what is likely to be generic, and what is largely an accident. In particular if alien life exists, it may share some Earthlike universals, but it's unlikely to share any of our parochials. Humanoid aliens, just like us, evolving independently on another world, would have too many parochials to be credible. Elbows, for instance. But aliens with limbs of some kind, able to move at will, exploit a universal.

Any specific alien design will be littered with parochials. If it's been constructed sensibly, it could be *similar* to an actual life form, somewhere with a similar environment. It would have suitable universals. But there's little chance that every single parochial would appear in the same real creature. Design a butterfly with fancy coloured wings, cute antennae, body markings... Now go and find a real one that's *exactly* the same. Not likely.

Since we're discussing the prospects for alien life, it seems sensible to ask what would count as 'life'. Specifying the meaning of 'life' too narrowly runs the risk of using parochials to define away highly complex entities that clearly ought to count as living. To avoid this danger, we should stick to universals. In particular, Earthlike biochemistry is *probably* a parochial. Experiments show that there can be innumerable viable variations on our familiar DNA/amino acid/protein system. If we encountered aliens who had developed a spacefaring civilisation, but didn't have DNA, it would be daft to insist that they're not alive.

I said 'specify' rather than 'define' because it's not clear that *defining* life makes sense. There are too many grey areas, and any form of words probably has an exception. Flames share many characteristics of life, including the ability to reproduce, but we wouldn't count them as living. Are viruses alive or not? The mistake is to imagine that there's a *thing* that we call life, and we have to pin down what that thing is. Life is a concept that our brains have extracted from the complexity of what's around us, and consider to be important. We get to choose what the word means.

Most of today's biologists today were trained in molecular biology, and by reflex they think in terms of organic (carbon-based) molecules. They've been extraordinarily clever to discover how life works on this planet, so it's hardly surprising if their default image of alien life looks much like life here. Mathematicians and physical scientists tend to think structurally. In this view, what matters about life, even on this planet, is not what it's made from. It's *how it behaves*.

One of the most general specifications of 'life' was invented by Stuart Kauffman, one of the founders of complexity theory. He uses a different term: autonomous agent. This is 'something that can both reproduce itself and do at least one thermodynamic work cycle'. Like all attempts, the intention is to capture key features that make living organisms special. It's not bad. It concentrates on behaviour, not ingredients. It avoids defining life away by focusing on its blurred boundaries, rather than recognising its remarkable differences from most other systems.

If we found something on another world that behaved like a computer program, we wouldn't declare it to be a form of alien life. We'd go looking for the creature that wrote it. But if we found something that satisfied Kauffman's conditions, I think we'd probably consider it to be alive.

✦

A case in point.

Some years ago, Cohen and I designed four alien environments for a museum project. The most exotic, which we named Nimbus, was loosely modelled on Titan. The original description had much more detail, such as evolutionary history and social structure.

Nimbus, as we envisaged it, is an exomoon with a dense atmosphere of methane and ammonia. A thick layer of cloud makes the surface very gloomy. The aliens of Nimbus are based on silicometallic chemistry, in which occasional metal atoms make it possible for silicon to form the backbone of large, complex molecules.[15] The metals come from meteorite impacts. Among the early life forms were metalloid mats of thin fibres carrying weak electric currents. They moved by putting out long tendrils. Small networks of tendrils could perform simple computations, and evolved to be more complex. These primitive creatures died out half a billion years ago, but they left a legacy, a silicon-based electronic ecology.

Today, the most striking visible features are fairy castles – intricate systems of roughly concentric silicometallic walls that retain ethane/methane pools. The pools are breeding grounds for flakes – electronic creatures that bootstrapped from the mats. Flakes are thin, flattish slivers of silica rock, overlaid with silicometallic electronic circuits. They are subject to complex evolutionary arms races in which they take over other flakes' circuitry. Every so often new circuits arise, better at taking over other circuits. By now they're quite good at this. The basis of their reproduction is template-copying. A mobile flake stamps a chemical image of its circuit on to virgin rock. This acts as a template for growing a mirror-image copy of the circuit. Then the copy splits off the rock. Copying errors permit mutations; taking over circuits leads to recombination of elements offering survival advantages in the arms race.

When humans discover Nimbus, some flakes are starting to go three-dimensional. They've become 'vonneumanns', replicating by a new trick. Around 1950 the mathematician John von Neumann introduced a cellular automaton (a simple kind of mathematical computer game) to prove that, in principle, self-replicating machines are possible.[16] It has three components: data, copier, and builder. The builder obeys coded instructions in the data to make a new builder and a new copier. Then the new copier copies the old data, and we have a second copy. A Nimbus vonneumann's circuitry is similarly segregated into three patches: data, copier, and builder. The builder can construct circuits prescribed by the data. The copier is just a copier. This ability has coevolved along with a three-sex reproductive system. One parent stamps a copy of its builder circuit on to bare rock. Later, another passes, notices the stamped circuit, and adds a copy of its copier. Finally, a third parent contributes a copy of its data. Now a new vonneumann can flake off.

'How different, how very different, from the home life of our own dear Queen,' as one of Queen Victoria's ladies-in-waiting reputedly said at a performance of Cleopatra by Sarah Bernhardt. No oxygen, no water, no carbon, no habitable zone, no genetics, three sexes... Complex enough to count as a form of life, albeit a highly unorthodox one, and able to evolve by natural selection. Yet the main features are scientifically realistic.

I'm not claiming that entities like this actually exist; in fact, no *specific* design of alien life is likely to exist, because it will involve too many parochials. But they illustrate the rich variety of new possibilities that might evolve on worlds very different from ours.

14

Dark Stars

HOLLY: Well, the thing about a black hole, its main distinguishing feature, is – it's black. And the thing about space, the colour of space, your basic space colour, is – it's black. So how are you supposed to see them?

Red Dwarf, Series 3 episode 2: 'Marooned'

FLYING TO THE MOON has long been a human dream. Lucian of Samosata's satirical *True Fictions*, dating from about 150 AD, includes imaginary trips to the Moon and Venus. In 1608 Kepler wrote a science fiction novel, *Somnium* (The Dream), in which demons convey a boy named Duracotus from Iceland to the Moon. In the late 1620s Francis Godwin, Bishop of Hereford, wrote *The Man in the Moone*, a rollicking tale in which gansas, giant swans, fly the sailor Domingo Gonsales to the Moon.

Kepler's demons were better science than Godwin's gansas. A swan, however powerful, can't fly to the Moon because space is a vacuum. But a demon can give sedated humans a big enough push to propel them off the planet. How big? The kinetic energy of a rocket is half its mass times the square of its velocity, and this must overcome the potential energy of whichever gravitational field it's attempting to escape from. Kepler was aware of this, though not in those words. In order to escape, the rocket must exceed a critical 'escape velocity'. Hurl something skywards faster, and it won't come back; slower, and it will. Earth's escape velocity is 11·2 kilometres per second. In the absence of any other bodies, and ignoring air resistance, that will give you a big enough impulse to escape the Earth forever. You still *feel* its gravitational force

– remember, law of *universal* gravitation – but the force falls away so rapidly that you don't grind to a halt. When other bodies are present, their combined effect must also be taken into account. If you start on Earth and want to escape from the Sun's gravity well, you need a velocity of 42·1 kilometres per second.

There are ways to get round this limit. The space bolas is a hypothetical device that swings a cabin round like one compartment on one spoke of a Ferris wheel. Fit several together in cascade and you can ride up to orbit on a series of spokes. Better still, you could build a space elevator – basically a strong rope hanging down from a satellite in geostationary orbit – and climb the rope as slowly as you wish. Escape velocity is irrelevant to these technologies. It applies to freely moving objects, given a big push and then left to their own devices. And that leads to a far more profound consequence of escape velocity, because one such object is the photon, a particle of light.

✦

When Rømer discovered that light has finite speed, a few scientists realised the implication: light can't escape from a sufficiently massive body. In 1783, John Michell imagined that the universe might be littered with huge bodies, bigger than stars but totally dark. In 1796 Laplace published the same idea in his masterwork *Exposition du Système du Monde* (Exposition of the System of the World):

> Rays from a luminous star having the same density as the Earth and a diameter 250 times that of the Sun would not reach us because of its gravitational attraction; it is therefore possible that the largest luminous bodies in the Universe may be invisible for this reason.

He deleted this passage from the third edition onwards, presumably because he was having doubts.

If so, he needn't have worried, although it took over two centuries to confirm the existence of his 'dark stars'. The Newtonian basis of the calculations had by then been overthrown by relativity, which put the dark star concept in a new light – or dark. Solutions of Einstein's field equations, for spacetime surrounding a very large dense mass, predict something even weirder than Michell's and Laplace's dark stars. Not only does such a mass trap all the light it emits; it disappears from the universe altogether, hidden behind a one-way ticket to oblivion called an event horizon. In 1964 the journalist Ann Ewing wrote an article about this idea with the catchy title 'black holes in space'. The physicist John Wheeler used the same term in 1967, and is often credited with inventing the name.

The *mathematical* existence of black holes is a direct consequence of general relativity, although some scientists wondered whether it exposes the theory as incomplete, lacking some additional physical principle to rule out such a bizarre phenomenon. The best way to resolve the issue is to observe a real black hole. This proved tricky, though not solely because of the memorable statement by the computer Holly in the British television programme *Red Dwarf*, quoted in the epigram above. Even if a black hole were invisible, its gravitational field would affect matter outside it in a characteristic way. Moreover (sorry, Holly), relativity implies that black holes aren't actually black, and they aren't exactly holes. Light can't get out, but the matter being sucked *in* produces effects that can be observed.

Today, black holes are no longer the stuff of science fiction. Most astronomers accept their existence. Indeed, it seems that most galaxies have a supermassive black hole at their centre. They may be why galaxies formed in the first place.

✦

The theory of black holes emerged from mathematical advances in general relativity, where matter warps spacetime and warped spacetime affects how matter moves, all in accordance with Einstein's field equations. A solution of the equations represents a possible geometry for spacetime, either in a limited region of the universe or for the universe as a whole. Unfortunately, the field equations are complicated – much more so than the equations of Newtonian mechanics, and even those are difficult enough. Before we had fast computers, the only way to find solutions to the field equations was by pencil, paper, and Hercule Poirot's 'little grey cells'. A useful mathematical trick in such circumstances is symmetry. If the required solution is spherically symmetric, the only variable that matters is the radius. So instead of the usual three dimensions of space, you have to consider only one, which is much easier.

In 1915 Karl Schwarzschild exploited this idea to solve the Einstein equations for the gravitational field of a massive sphere, modelling a large star. The reduction to one space variable simplified the equations enough for him to derive an explicit formula for the geometry of spacetime around such a sphere. At the time he was in the Prussian army, fighting the Russians, but he managed to send his discovery to Einstein, asking him to sort out publication. Einstein was impressed, but Schwarzschild died six months later of an incurable autoimmune disease.

One of the common delights of mathematical physics is that equations often seem to know more than their creators do. You set up equations based on physical principles that you understand pretty well. Then you bash out a solution, find out what it tells you, and discover that you don't understand the answer. More precisely, you understand what the answer *is*, and why it solves the equations, but you don't fully understand why the answer behaves as it does.

That, by the way, is what equations are *for*. If we could always guess the answers in advance, we wouldn't need the equations. Think of Newton's law of gravity. Can you look at the formula and see an ellipse? I know I can't.

Anyway, Schwarzschild's results contained a big surprise: his solution behaved very strangely at a critical distance, now called the Schwarzschild radius. In fact, the solution has a singularity there: some terms in the formula become infinite. Inside a sphere of this critical radius, the solution tells us nothing sensible about space or time.

The Sun's Schwarzschild radius is three kilometres, and the Earth's is a mere centimetre – both buried inaccessibly deep where they can't cause trouble, but also inaccessible to observations, making it difficult to compare Schwarzschild's answer to reality, or to find out what it means. This weird behaviour raised a basic question: what would happen for a star so dense that it lies inside its own Schwarzschild radius?

✦

Leading physicists and mathematicians got together in 1922 to argue about this question, but came to no clear conclusion. The general feeling was that such a star will collapse under its own gravitational attraction. What happens then depends on the detailed physics, and at the time this was mainly guesswork. By 1939 Robert Oppenheimer had calculated that sufficiently massive stars will indeed undergo gravitational collapse in such circumstances, but he believed that the Schwarzschild radius bounds a region of spacetime in which time comes to a complete stop. This led to the name 'frozen star'. However, this interpretation was based on an incorrect assumption about the region of validity of the Schwarzschild solution. Namely, that the singularity has a genuine physical meaning. From the viewpoint of an external observer, time does indeed stop at the Schwarzschild radius. However, that's not true for an observer who falls in past the singularity. This duality of viewpoints runs like a golden thread through the theory of black holes.

In 1924 Arthur Eddington showed that Schwarzschild's singularity is a mathematical artefact, not a physical phenomenon. Mathematicians represent curved spaces and spacetimes using a mesh of curves or surfaces labelled by numbers, like lines of latitude and longitude on the Earth. These meshes are called coordinate systems. Eddington showed that Schwarzschild's singularity is a special feature of his choice of coordinates. Analogously, all lines of longitude meet at the North Pole, and lines of latitude there form ever smaller circles. But if you stand at the North Pole, the surface looks the same *geometrically* as it does anywhere else. Just more snow and ice. The apparently strange geometry near the North Pole is caused by choosing latitude and longitude as coordinates. If you used a coordinate system with an East Pole and a West Pole on the equator, those points would seem weird, and the North and South Poles would seem normal.

Schwarzschild's coordinates represent what the black hole looks like from outside, but from inside it looks quite different. Eddington found a new system of coordinates that makes the Schwarzschild singularity disappear. Unfortunately, he failed to follow up this discovery because he was working on other astronomical questions, so it went largely unnoticed. It became better known in 1933 when Georges Lemaître independently realised that the singularity in Schwarzschild's solution is a mathematical artefact.

Even then, the topic languished until 1958, when David Finkelstein found a new improved coordinate system in which the Schwarzschild radius has physical meaning – but not that time freezes there. He used his coordinates to solve the field equations not only for an external observer, but for the entire future of an internal observer. In these coordinates there's no singularity at the Schwarzschild radius. Instead, it constitutes an *event horizon*: a one-way barrier whose outside can influence its inside, but not the other way round. His solution demonstrates that a star lying inside its Schwarzschild radius collapses to form a region of spacetime from which no matter, not even photons, can escape. Such a region is partially disconnected from the rest of the universe – you can get into it, but you can't get back out. This is a true black hole in the current sense of the term.

What a black hole looks like depends on the observer. Imagine an unfortunate spaceship – well, a spaceship whose crew is unfortunate – falling into a black hole. This is a staple of science fiction movies, but very few of them get it remotely right. The movie *Interstellar* did, thanks to advice from Kip Thorne, but its plot has other defects. The physics shows that if we watch the falling spaceship from a distance, it seems to move ever more slowly, because the black hole's gravity tugs ever more strongly at the photons emitted by the spaceship. Those sufficiently close to the black hole can't escape at all; those just outside the event horizon, where gravity exactly cancels the speed of light, can escape, but very slowly. We observe the spaceship by detecting the light it emits, so we see it crawling to a halt, never reaching the event horizon. General relativity tells us that gravity slows time down. At the Schwarzschild radius, time *stops* – but only as seen by an external observer. The hole itself also becomes redder and redder, thanks to the Doppler effect. Which is why black holes are not, despite Holly's trademark sarcasm, black.

The crew of the spaceship experience none of this. They plunge towards the black hole, are sucked in across the event horizon, and then –

– they experience the solution of the equations as observed from *inside* the black hole. Possibly. We don't know for sure, because the equations say that all of the matter in the spaceship will be compressed into a single mathematical point of infinite density and zero size. This, if it actually happened, would be a genuine physical singularity, not to mention fatal.

Mathematical physicists are always a bit shy about singularities. Usually, when a singularity pops up, it means that the mathematical model is losing contact with reality. In this case we can't send a probe into a black hole and bring it out again, or even receive radio signals from it (which travel at the speed of light and also can't escape), so there's no way to find out what the reality is. However, it seems likely that, whatever happened, it would be unpleasantly violent and the crew wouldn't survive. Except in movies. Well, some crew in some movies.

✦

The mathematics of black holes is subtle, and initially the only type of black hole for which the field equations could be solved explicitly was Finkelstein's, which applies to a black hole that doesn't rotate and has no electric field. This type is often called a Schwarzschild black hole. The mathematical physicist Martin Kruskal had already found a similar solution, but not published it. Kruskal and George Szekeres developed this into what are now called Kruskal–Szekeres coordinates, which describe the interior of the black hole in more detail. The basic geometry is very simple: a spherical event horizon with a point singularity at its centre. Everything that falls into the black hole reaches the singularity in a finite period of time.

This kind of black hole is special, because most celestial bodies spin. When a spinning star collapses, conservation of angular momentum requires the resulting black hole to spin as well. In 1963 Roy Kerr produced a mathematical rabbit from a hat by writing down a spacetime metric for a rotating black hole – the Kerr metric. Since the field equations are nonlinear, an explicit formula is remarkable. It

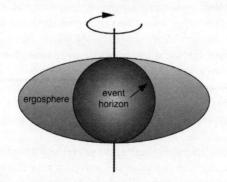

Event horizon (sphere) and ergosphere (ellipsoid) for a spinning black hole.

shows that, instead of a single spherical event horizon, there are two critical surfaces at which physical properties change dramatically. The innermost is a spherical event horizon; just as for the static black hole it represents a barrier that light can't cross. The outer one is a flattened ellipsoid, touching the event horizon at the poles.

The region between is called the ergosphere. *Ergon* is Greek for 'work', and the name arises because you can extract energy from the black hole by exploiting the ergosphere. If a particle falls inside the ergosphere, a relativistic effect called frame-dragging makes it start to spin along with the black hole, which increases its energy. But since the particle is still outside the event horizon, it can – in suitable circumstances – escape, taking that energy with it. It thereby extracts energy, something you can't do with a static black hole.

As well as spin, a black hole can have electrical charge. Hans Reissner and Gunnar Nordström found a metric for a charged black hole: the Reissner–Nordström metric. In 1965 Ezra Newman discovered the metric for an axisymmetric rotating charged black hole, the Kerr–Newman metric. You might think that even more elaborate types of black hole could exist, but physicists believe not, except possibly a magnetic one. The no-hair conjecture asserts that once the black hole has settled down after its initial collapse, and ignoring quantum effects, it has only three basic physical properties: mass, spin, and charge. The name comes from the phrase 'black holes have no hair' in the 1973 bible of the subject: *Gravitation* by Charles Misner, Kip Thorne, and John Wheeler. Wheeler has attributed the phrase to Jacob Bekenstein.

This statement is often called the no-hair *theorem*, but it has not yet been proved, which is what that word normally implies. Neither has it been disproved, for that matter. Stephen Hawking, Brandon Carter, and David Robinson have proved some special cases. If, as some physicists think, a black hole can also have a magnetic field, the conjecture would have to be modified to include that possibility too.

✦

Let's do some black hole geometry, to get a feel for how strange these structures are.

In 1907 Hermann Minkowski devised a simple geometric picture of relativistic spacetime. I'll use a simplified image with only one dimension of space plus the usual one of time, but this can be extended to the physically realistic case with three dimensions of space. In this representation, curved 'worldlines' represent the motion of particles. As the time coordinate changes, you can read off the resulting space coordinate from the curve. Lines at an angle of 45 degrees to the axes represent particles moving at the speed of light. Worldlines therefore can't cross any 45-degree line. A point in spacetime, called an event, determines two such lines, which together form its light cone. This comprises two triangles: its past and its future. The rest of spacetime is inaccessible starting from that point – to get there, you'd have to travel faster than light.

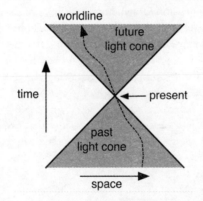

Minkowski's representation of relativistic spacetime.

In Euclid's geometry, the natural transformations are rigid motions, and these preserve *distances* between points. The analogues in special relativity are Lorentz transformations, and these preserve a quantity called the interval. By Pythagoras's theorem the square of the distance from the origin to a point in a plane is the sum of the squares of horizontal and vertical coordinates. The square of the interval is the square of the space coordinate *minus* that of the time coordinate.[1] This difference is zero along the 45-degree lines, and positive inside the light cone. So the interval between two causally connected events is a real number. Otherwise it's an imaginary number, reflecting the impossibility of travelling between them.

In general relativity, gravity is included by allowing Minkowski's flat plane to bend, mimicking the effects of a gravitational force, as in the picture on page 25.

Recasting Minkowski's geometry in Kruskal–Szekeres coordinates, Roger Penrose developed a beautifully simple way to picture the relativistic geometry of black holes.[2] The formula for the metric implicitly determines this geometry, but you can stare at the formula until you're blue in the face without getting anywhere. Since it's geometry we want, how about drawing pictures? The pictures must be consistent with the metric, but a good picture is worth a thousand calculations.

Penrose diagrams reveal subtle features of black hole physics, allowing comparisons between different kinds of black hole. They also lead to some startling, albeit speculative, possibilities. Again space is reduced to one dimension (drawn horizontally), time is drawn vertically, and light rays travel at 45-degree angles to form light cones separating past, future, and causally inaccessible regions.

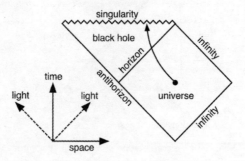

Penrose diagram of Schwarzschild black hole.

The Minkowski picture is normally drawn as a square, but Penrose diagrams use a diamond shape instead, to emphasise the special nature of 45-degree slopes. Both shapes are just different ways to compress an infinite plane into a finite space. They're unusual but useful coordinate systems on spacetime.

As a warm-up, let's start with the simplest kind, the Schwarzschild black hole. Its Penrose diagram is fairly simple. The diamond represents the universe, essentially following the Minkowski model. The arrowed curve is the worldline of a spacecraft falling into the black hole by crossing its (event) horizon, and hitting the central singularity (zigzag line). But now there's a second horizon, labelled 'antihorizon'. What's that about?

When we discussed a spacecraft falling into a black hole, we discovered that this process looks very different if you're inside the spacecraft or watching from outside the black hole. The spacecraft follows a path like the curved arrow in the picture, travelling across the horizon and into the singularity. But because light escapes ever more slowly as the spacecraft approaches the horizon, an external observer sees an ever redder spacecraft, slowing down and coming to an apparent halt. The colour change is caused by gravitational redshift: gravitational fields slow time down, changing the frequency of an electromagnetic wave. Other objects that had fallen in would also be visible, every time anyone looked. Once frozen, they would appear to stay that way.

The horizon in the Penrose diagram is the event horizon as observed by the crew. The antihorizon is where the spacecraft *appears* to come to rest, as seen by an external observer.

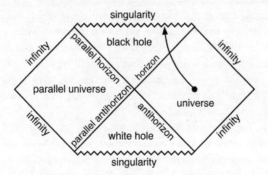

Penrose diagram of Schwarzschild black hole/white hole pair.

A curious mathematical construction now becomes possible. Suppose we ask: what lies on the far side of the antihorizon? In the reference frame of the crew, it's the interior of the black hole. But there's a natural mathematical extension of Schwarzschild geometry, in which a time-reversed copy of a Schwarzschild black hole is glued on to an ordinary one. Mathematically we glue two copies of the metric together, reversing time in one by rotating the picture through 180 degrees, to get the complete picture.

A time-reversed black hole is known as a white hole, and it behaves like a black hole when time runs backwards. For a black hole, stuff (and light) falls in but can't get out. For a white hole, stuff (and light) falls out but can't get in. A 'parallel horizon' emits light and matter, but is impermeable to either if it tries to enter the white hole.

The rotated image of our universe also describes a universe, but it's not causally connected to ours, because relativity's lightspeed limit implies that you can't get inside it by following a path that's steeper than 45 degrees. Speculating, the second image could represent a different universe altogether. Entering the realm of pure fantasy, technology sufficiently advanced to allow faster-than-light travel could transit between these two universes, while avoiding the singularities.

If a white hole is connected to a black hole, in a way that allows light, matter, and causal effects to pass through, we get a 'wormhole', much beloved of science fiction books and movies as a way to overcome the cosmic speed limit and get the characters to an alien planet before they die of old age. A wormhole is a cosmic shortcut between different universes, or different regions of the same universe. Since everything that enters a black hole is preserved as a frozen image when viewed by an outside observer, a wormhole that's been in regular use will appear to be surrounded by frozen, reddened images of every vessel that has entered its black hole mouth. I've not seen that in any science fiction movie.

In this case, the black and white holes are not connected like that, but in the next type of black hole they are. This is the rotating or Kerr black hole, and it's bizarre. Start from a Schwarzschild black hole/ white hole pair, but without the singularity. Extend both black and white hole regions to form diamonds. Between these diamonds insert (on the left) a new diamond. This has a *vertical* singularity (fixed in space but persisting over time). On one side (the right-hand one in the Penrose diagram) of the singularity is a 'wormhole' region that links the black and white holes while avoiding the singularity. Following the wiggly path through the wormhole leads from this universe to a new one. On the other side (left) of the singularity is an antiverse: a universe filled with antimatter. Similarly, add another diamond on the right representing a parallel wormhole and antiverse.

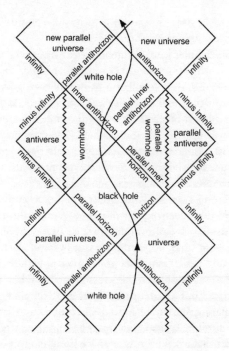

Penrose diagram of a rotating (Kerr) black hole.

But this is just the start. Now make an infinite stack, tiled by these diamonds. This construction 'unwraps' the spin of the black hole, and produces an infinite sequence of wormholes linking infinitely many different universes.

Geometrically, the singularity of a Kerr black hole isn't a point, but a circular ring. By passing through the ring, it's possible to travel between a universe and an antiverse. Though perhaps unwise, given what antimatter does to matter.

The Penrose diagram for a charged (Reissner–Nordström) black hole is similarly elaborate, with some differences of interpretation. The mathematics doesn't imply that all of these strange phenomena actually exist or occur. It implies that they're natural consequences of the mathematical structure of a rotating black hole – structures for spacetime that are logically consistent with known physics, hence reasonable consequences of it.[3]

✦

That's what black holes look like geometrically, but how can they arise in reality?

A massive star starts to collapse under its own gravity when the nuclear reactions that make it shine run out of fuel. If so, how does the mattter in the star behave? That's a far more complicated problem today than it was for Michell and Laplace. The stars haven't changed, but our understanding of matter has. Not only must we think about gravity (and use relativity, not Newton); we must also consider the quantum mechanics of nuclear reactions.

If a lot of atoms are forced ever closer together by gravity, their outer regions, occupied by electrons, try to merge. A quantum-theoretic fact, the Pauli exclusion principle, implies that no two electrons can occupy the same quantum state. So, as the pressure grows, electrons seek out any unoccupied states. Soon they're jammed together cheek-by-jowl like a pile of oranges outside the greengrocer's. When the electrons run out of space, and all quantum states are occupied, they become electron degenerate matter. This occurs in the cores of stars.

In 1931 Subrahmanyan Chandrasekhar used relativistic calculations to predict that a sufficiently massive body, composed of electron degenerate matter, must collapse under its own gravitational field to form a neutron star, composed almost entirely of neutrons. A typical neutron star manages to compress twice the mass of the Sun into a sphere with a 12 kilometre radius. If the mass is less than 1·44 times that of the Sun, a figure called the Chandrasekhar limit, it forms a white dwarf instead of a neutron star. If it's more massive, but below the Tolman–Oppenheimer–Volkoff limit of three times the solar mass, it collapses to form a neutron star. At that stage, further collapse to a black hole is to some extent prevented by neutron degeneracy pressure, and astrophysicists are unsure of the outcome. However, anything with a mass more than ten times that of the Sun will overcome the pressure and become a black hole. The smallest mass yet observed in a black hole is about five times that of the Sun.

A purely relativistic model indicates that a black hole itself can't emit radiation – only the matter being sucked in can do that, being outside the event horizon. But Hawking realised that quantum effects can cause a black hole to emit radiation from its event horizon.

Quantum mechanics permits the spontaneous creation of a virtual particle–antiparticle pair, as long as they annihilate each other very shortly afterwards. Or would do, except that when this happens just outside the event horizon, the black holes's gravity pulls one particle across the event horizon and (by conservation of momentum) leaves the other outside, whence it can escape altogether. This is Hawking radiation, and it makes small black holes evaporate very quickly. Big ones evaporate too, but the process takes a huge amount of time.

✦

The Einstein field equations have mathematical black hole solutions, but that's no guarantee that they occur in nature. Perhaps unknown laws of physics prevent black holes. So before we get too carried away with the mathematics and astrophysics, it's a good idea to find observational evidence that black holes exist. It would be fascinating to go further, looking for white holes and wormholes and alternative universes, but right now black holes are ambitious enough.

Initially, black holes remained theoretical speculation, impossible to observe directly because the only radiation they emit is weak Hawking radiation. Their existence is deduced indirectly, mainly from gravitational interactions with other nearby bodies. In 1964 an instrument on a rocket spotted an exceptionally strong X-ray source in the constellation Cygnus, known as Cygnus X-1. Cygnus, the swan, flies along the Milky Way, which is significant because Cygnus X-1 lies at the heart of our Galaxy, and therefore appears to us to be somewhere inside the Milky Way.

In 1972 Charles Bolton, Louise Webster, and Paul Murdin combined observations from optical and radio telescopes to show that Cygnus X-1 is a binary object.[4] One component, which emits visible light, is the blue supergiant star HDE 226868. The other, detected only by its radio emissions, is about 15 times as massive as the Sun but so compact that it can't be any normal kind of star. Its estimated mass exceeds the Tolman–Oppenheimer–Volkoff limit, so it isn't a neutron star. This evidence made it the first serious candidate for a black hole. However, the blue supergiant is so massive that it's difficult to estimate the mass of the compact component accurately. In 1975 Thorne and Hawking had a bet about it: Thorne said it was a black hole and Hawking said it wasn't. After extra observations in 1990 Hawking conceded defeat and paid up, although the object's status is still not definitively confirmed.

More promising X-ray binaries exist for which the ordinary component is less massive. The best of these is V404 Cygni, discovered in 1989 and now known to be 7800 light years away. Here the ordinary component is a star slightly smaller than the Sun, and the compact component has mass about 12 times that of the Sun, above the Tolman–Oppenheimer–Volkoff limit. There's other supporting evidence, so it's generally agreed to be a black hole. The two bodies orbit every six and a half days. The black hole's gravity deforms the star into an egg shape, stealing its material in a steady stream. In 2015 V404 began emitting short bursts of light and intense X-rays, something that had previously occurred in 1938, 1956, and 1989. It's thought that the cause is material that piles up around the black hole and is sucked in when its mass exceeds a critical value.

Other black holes have been detected through the X-rays they emit. Infalling gas forms a thin disc called an accretion disc, and the gas is then heated by friction as angular momentum migrates outwards through the disc. The gas can heat up so much that it produces highly energetic X-rays, and up to 40% of it can be turned into radiation. Often the energy is carried away in enormous jets at right angles to the accretion disc.

A fascinating recent discovery is that most sufficiently large galaxies have a giant central black hole, massing between 100,000 and a billion Suns. These supermassive black holes may organise matter into galaxies. Our Galaxy has one, the radio source Sagittarius A*. In 1971 Donald Lynden-Bell and Martin Rees presciently suggested that this might be a supermassive black hole. In 2005 it was discovered that M31, the Andromeda Galaxy, has a central black hole with a mass of 110–230 million Suns. Another galaxy in our neighbourhood, M87, has a black hole whose mass is 6·4 billion Suns. The distant elliptical radio galaxy 0402+379 has two supermassive black holes that orbit each other like a gigantic binary star system, 24 light years apart. It takes 150,000 years for them to complete one orbit.

✦

Most astronomers accept that such observations show the existence of

black holes in the conventional relativistic sense, but there's no definitive evidence that this explanation is correct. At best it's circumstantial,
based on current theories of fundamental physics, even though we
know that relativity and quantum mechanics are uneasy bedfellows
– especially when, as here, we need to invoke both at the same time.
A few maverick cosmologists are starting to wonder whether what
we see are *really* black holes, or something else that looks much the
same. They're also wondering whether our theoretical understanding
of black holes needs a rethink.

According to Samir Mathur, *Interstellar* doesn't work. You can't fall
into a black hole. We've seen that, contrary to what was first thought,
black holes can emit radiation for quantum reasons. This is Hawking
radiation, in which one particle from a transient virtual particle/anti-
particle pair falls into the black hole, while the other escapes. That
leads to the black hole information paradox: information, like energy,
is conserved, so it can't fall out of the universe. Mathur resolves the
paradox by presenting a different view of a black hole: a fuzzball that
you can stick to, but not penetrate.

In this theory, when you hit a black hole you don't fall in. Instead,
your information is spread thinly over the event horizon and you turn
into a hologram. It's not a new idea, but the latest version allows the
hologram to be an imperfect copy of the infalling object. This proposal
is controversial, in part because the same logic seems to demonstrate
that an event horizon is a highly energetic firewall, and anything hitting
it will roast. Fuzzball or firewall? The question is moot. Possibly both
are artefacts of an inapproriate coordinate system, like the discredited
view that an event horizon freezes time. On the other hand, we can't
distinguish what an outside observer sees from what an infalling one
sees, if nothing can fall in.

In 2002 Emil Mottola and Pawel Mazur challenged the prevailing
wisdom on collapsing stars. Instead of becoming a black hole, they
suggested that it might turn into a gravastar – a hypothetical strange
bubble of very dense matter.[5] From outside, a gravastar should look
much like a conventional black hole. But its analogue of the event
horizon is actually a cold, dense shell inside which space is springy.
The proposal remains controversial and several tricky issues are unresolved – such as precisely how such a thing forms – but it's intriguing.

The theory came from re-examining the relativistic scenario for a black hole in the light of quantum mechanics. The usual treatment ignores these effects, but that leads to strange anomalies. The information content of a black hole, for example, is far greater than that of the star that collapsed – but information ought to be conserved. A photon falling into a black hole should acquire an infinite amount of energy by the time it encounters the central singularity.

Mottola and Mazur, puzzled by these problems, wondered whether a proper quantum treatment could resolve them. When a collapsing star gets close to forming an event horizon, it creates a huge gravitational field. This distorts the quantum fluctuations of spacetime, leading to a different kind of quantum state akin to a gigantic 'superatom' (jargon: Bose–Einstein condensate). This is a cluster of identical atoms in the same quantum state, at a temperature close to absolute zero. The event horizon would become a thin shell of gravitational energy, like a shockwave in spacetime. This shell exerts negative pressure (that is, in the outward direction), so matter that falls inside it will turn round and rise back up to hit the shell. Matter from outside would still be sucked in, however.

Gravastars make mathematical sense: they're stable solutions of the Einstein field equations. They avoid the information paradox. Physically, they differ markedly from black holes, yet they look the same from outside: the external Schwarzschild metric. Suppose a star with 50 times the mass of the Sun collapses. Conventionally, you get a black hole 300 km across, which would emit Hawking radiation. In the alternative theory, you get a gravastar if the same size, but its shell is a mere 10^{-35} metres thick, its temperature is 10 billionths of a kelvin, and it emits no radiation whatsoever. (Holly would be pleased.)

Gravastars are a possible explanation of another baffling phenomenon: gamma-ray bursters. Every so often the sky lights up with a flash of high-energy gamma rays. The usual theory is that these are colliding neutron stars, or black holes formed during a supernova. The birth of a gravastar is another possibility. More speculatively, the interior of a gravastar the size of our universe would also be subject to negative pressure, which would accelerate matter towards its event horizon – that is, away from the centre. The sums suggest that this would be about the same size as the accelerated expansion of the universe normally attributed to dark energy. Perhaps our universe is actually the inside of a huge gravastar.

Among Einstein's predictions, over a century ago, was the occurrence of gravitational waves, which create ripples in spacetime like those on a pond. If two massive bodies, such as black holes, spiral rapidly round each other, they stir up the cosmic pond and create detectable ripples. In February 2016 the Laser Interferometer Gravitational-Wave Observatory announced the detection of gravitational waves caused by two merging black holes. LIGO's instruments are pairs of 4-kilometre tubes in an L shape. Laser beams bounce back and forth along the tubes, and their wave patterns interfere with each other at the junction of the L. If a gravitational wave passes by, the lengths of the tubes change slightly, affecting the interference pattern. The apparatus can detect a movement one thousandth of the width of a proton.

The signal that LIGO picked up matches the relativistic prediction for a spiralling collision between two black holes, with masses 29 and 36 times that of the Sun. The feat opens up a new era in astronomy: LIGO is the first successful graviscope, observing the cosmos using gravity instead of light.

This remarkable gravitational discovery provides no information on the more contentious quantum features that distinguish conventional black holes from hypothetical alternatives such as fuzzballs, firewalls, and gravastars. Its successors, floating in space, will be able to spot not just black hole collisions, but less violent mergers of neutron stars, and should help to resolve these mysteries. Meanwhile, LIGO has turned up a new mystery: a brief burst of gamma rays apparently related to the gravitational wave. Prevailing theories of merging black holes don't predict that.

We've got used to the existence of black holes, but they occupy a realm where relativity and quantum theory overlap and clash. We don't really know what physics to use, so cosmologists try their best with what's available. The last word on black holes is not yet in, and there's no reason to suppose that our current understanding is complete – or correct.

Skeins and Voids

The heaven, moreover, must be a sphere, for this is the only form worthy of its essence, as it holds the first place in nature.

Aristotle, *On the Heavens*

WHAT DOES THE UNIVERSE LOOK LIKE? How big is it? What shape is it?

We know something about the first question, and it's not what most astronomers or physicists originally expected. On the largest scales we can observe, the universe is like the foam in a washing-up bowl. The bubbles in the foam are gigantic voids, empty of almost all matter. The soap films surrounding the bubbles are where the stars and galaxies congregate.

Embarrassingly, our favourite mathematical model of the spatial structure of the universe assumes that matter is smoothly distributed. Cosmologists console themselves that on even larger scales, individual bubbles cease to be discernible and foam looks pretty smooth – but we don't know that the matter in the universe behaves like that. So far, every time we've observed the universe on a bigger scale, we've found bigger clumps and voids. Perhaps the universe isn't smooth at all. Perhaps it's a fractal – a shape with detailed structure on all scales.

We also have some ideas about the second question, size. The stars are not a hemispherical bowl erected over the Earth, as some ancient civilisations believed and *Genesis* seems to assume. They're a doorway into a universe so vast that it seems infinite. Indeed, it may *be* infinite. Many cosmologists think so, but it's difficult to imagine how to test that claim scientifically. We have a fair idea of how big the observable universe is, but beyond that, how can we even start to find out?

The third question, shape, is even trickier. Right now there's no agreement on the answer, although the consensus is the most boring contender, a sphere. There has long been a tendency to assume that the universe is spherical, the interior of a huge ball of space and matter. But at various recent times it has also been thought to be a spiral, a doughnut, a football, or a non-Euclidean geometric shape called the Picard horn. It might be flat or curved. If so, its curvature might be positive or negative, or vary from place to place. It might be finite or infinite, simply connected or full of holes – or even disconnected, falling into separate pieces that can never interact with each other.

✦

Most of the universe is empty space, but there's also plenty of matter – some 200 billion galaxies, with 200–400 billion stars in each. The way the matter is distributed – how much there is in any given region – is important, because Einstein's field equations relate the geometry of spacetime to the distribution of matter.

The matter in the universe is definitely *not* smeared out uniformly on the scales we've observed, but this discovery dates back only a few decades. Before that, the general view was that above the scale of galaxies, the overall distribution of matter looks smooth, much as a lawn looks smooth unless you can see individual blades of grass. But our universe seems more like a lawn with big patches of clover and mud, which create new non-smooth structure on larger scales. And when you try to smooth those out by taking an even broader view, the lawn disappears and you see the supermarket car park. More prosaically, there's a distinct tendency for the cosmic distribution of matter to be clumpy on a huge range of scales.

In our own neighbourhood, most of the matter in the solar system has clumped together to form a star, the Sun. There are smaller bits, planets, and even smaller ones – moons, asteroids, Kuiper belt objects … plus assorted small rocks, pebbles, dust, molecules, atoms, photons. Going in the opposite direction, to larger scales, we find other types of clumping. Several stars may be gravitationally bound to form a binary or multiple star system. Open clusters are groups of a thousand or so stars that all formed at much the same time in the same collapsing molecular cloud. They occur inside galaxies; about 1100 of them are known in our own Galaxy. Globular clusters are composed of hundreds of thousands of old stars in a huge fuzzy spherical ball, and they generally occur as orbiting satellites to galaxies. The Galaxy has 152 known ones and perhaps 180 altogether.

Galaxies are striking examples of the clumpiness of the universe: blobs, discs, and spirals that contain between a thousand and 100 trillion stars, with diameters ranging from 3000 to 300,000 light years. But galaxies aren't spread out uniformly either. They tend to occur in closely associated groups of about fifty, or in larger numbers (up to a thousand or so) as galactic clusters. The clusters in turn aggregate to form superclusters, which clump together into unimaginably vast sheets and filaments, with colossal voids in between.

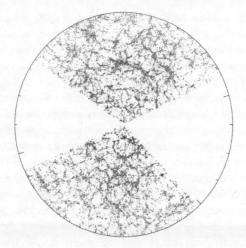

Two slices through the Sloan Digital Sky Survey of galaxies, showing filaments and voids. Earth has been placed at the centre. Each dot is a galaxy, and the radius of the circle is 2 billion light years.

We, for instance, are in the Galaxy, which is part of the *local group* of galaxies, along with the Andromeda Galaxy M31 and another 52 galaxies, many of them dwarf galaxies like the Magellanic Clouds that act as satellites to the two dominant spirals: Andromeda and us. About ten dwarf galaxies are not gravitationally bound to the rest. The other main large galaxy in the local group is the Triangular Galaxy, which may be a satellite of Andromeda. The whole local group is about ten million light years across. The local group is part of the Laniakea Supercluster, identified in 2014 in an exercise to define superclusters mathematically by analysing how fast galaxies move relative to one another. The Laniakea Supercluster is 520 million light years across and contains 100,000 galaxies.

As new and ever larger clumps and voids are found, cosmologists keep revising the scale on which they think the universe should be smooth. The current view is that clumps and voids should be no larger than a billion light years, and most should be smaller. Some recent observations are therefore rather disconcerting. A team under András Kovács has found a void 2 billion light years across, and Roger Clowes and coworkers have found a coherent cosmological structure twice that size, the Huge Large Quasar Group, which contains 73 quasars. These objects are respectively twice and four times the largest expected size for a unified structure. Lajos Balász's group has observed a ring of gamma-ray bursters 5·6 billion light years in diameter, which is even bigger.[1]

These discoveries, and even more so their explanations, are controversial. Some dispute the meaning of the observations. Others argue that a few unusually large structures don't stop the universe being homogeneous 'on average'. That's true, but not entirely convincing, because these structures simply do not fit the standard mathematical model: a manifold that is smooth not just on average, but everywhere – except for deviations less than a billion light years across. All previous assertions of smoothness on smaller scales have come to pieces as new, more far-reaching surveys have been carried out. It seems to be happening again.

Identifying clusters, incidentally, is a non-trivial task. Just what counts as a cluster or supercluster? The human eye naturally sees clumps, but these need not be meaningfully related in gravitational terms. The solution uses a mathematical technique called Wiener filtering, an elaborate type of least-squares data-fitting that can separate signals from noise. Here it's adapted to separate the movements of galaxies into a part representing the expansion of the universe, which is common to all galaxies, and another part that gives their individual 'proper motions' relative to that expansion. Galaxies in the same general region that have similar proper motions belong to the same supercluster. The cosmos is like a fluid, with stars for atoms, galaxies for vortices, and superclusters for larger-scale structures. Using Wiener filtering, you can work out the flow patterns of this fluid.

Cosmologists have simulated how matter in the universe clumps together under gravity. The general picture of thin skeins and sheets of matter separated by gigantic voids seems to be a natural structure for a large system of bodies interacting through gravity. But it's much harder to make the statistics of the skeins and sheets match observations, or to obtain a realistic distribution of matter within the orthodox timescale of 13·8 billion years.

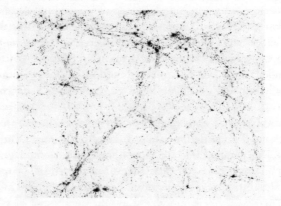

Computer simulation of a region 50 million light years across for one model of the distribution of visible matter in the universe.

The usual way to get round this is to invoke the existence of esoteric particles called dark matter. This assumption in effect strengthens the force of gravity, so large structures can evolve more rapidly, but it's not entirely satisfactory, see Chapter 18. An alternative, largely ignored, is the possibility that the universe is a lot older than we think. A third is that we haven't yet hit on the right model.

✦

Size next.

As astronomers probed the universe with ever more powerful tele-scopes, they weren't just seeing further; they were looking back in time. Since light has a finite speed, it takes a definite amount of time to travel from one place to another. Indeed, the light year is defined to be the distance that light travels in a year.

Light moves very fast, so a light year is a very long distance: 9·46 trillion kilometres. The nearest star is 4·24 light years away, so when anyone views it in a telescope they are seeing what it was like four and a quarter years ago. For all we know, it might have exploded yesterday (unlikely, mind you: it's not at that stage in its evolution), but if it did, we won't be able to find out for another four and a quarter years.[2]

The current figure for the radius of the observable universe is about 45·7 billion light years. Naively, we might imagine that we can therefore see 45·7 billion years into the past. However, we can't, for two reasons. First, 'observable universe' refers to what it would be possible to observe in principle, not what we can see in practice. Second, the universe is currently believed to be only 13·8 billion years old. The missing 31·9 billion years is accounted for by the expansion of the universe, but I'll come back to that in the next chapter.

That's an awful lot of universe. And that's just the observable part. There could be more. I'll come back to that, too. At any rate, we can give a well-informed answer to the question 'how big is the universe?' if we interpret it in a reasonable way.

✦

In contrast, the question 'what shape is the universe?' is much harder to answer, and the source of much debate.

Before Einstein worked out how to to incorporate gravity into his relativistic theory of spacetime, nearly everyone assumed that the geometry of space must be that of Euclid. One reason was that for a large part of the time between Euclid writing the *Elements* and Einstein's radical revision of physics, it was generally believed that the only possible geometry is Euclid's.

That belief was torpedoed in the 1800s when mathematicians discovered self-consistent non-Euclidean geometries, but although these have beautiful applications within mathematics, hardly anyone expected them to apply to the real world. An exception was Gauss, who discovered non-Euclidean geometry but kept it quiet because he doubted anyone would accept it, and he preferred to avoid uninformed criticism. Of course, people knew about geometry on the surface of a sphere; navigators and astronomers routinely used an elaborate theory of spherical trigonometry. But that was all right because a sphere is just a special surface in ordinary Euclidean space. It's not space itself.

It occurred to Gauss that if geometry need not be Euclidean, real space need not be Euclidean either. One way to distinguish the various geometries is to add up the angles of a triangle. In Euclidean geometry you always get 180 degrees. In one type of non-Euclidean geometry, elliptic geometry, it's always more than 180 degrees; in the other type, hyperbolic geometry, it's always less than 180 degrees. The precise number depends on the area of the triangle. Gauss tried to find out the true shape of space by measuring a triangle formed by three mountain peaks, but he didn't get a convincing result. Ironically, in view of what Einstein did with the mathematics that emerged from these discoveries, the gravitational attraction of the mountains interfered with his measurements.

Gauss started wondering how to quantify the curvature of a surface: how sharply it bends. Up till then, a surface was traditionally viewed as the boundary of a solid object in Euclidean space. Not so, said Gauss. You don't need the solid object: the surface alone is enough. You don't need a surrounding Euclidean space either. All you need is something that determines the surface, and in his view that something is a notion of distance, a *metric*. Mathematically, a metric is a formula for the distance between any two points that are very close together. From this, you can work out how far apart any two points are, by stringing together a series of very close neighbours, using the formula to find out how far apart those are, adding up all those small distances, and then choosing the string of neighbours to make the result as small as possible. The string of neighbours fits together to make a curve called a geodesic, and it's the shortest path between those two points. This idea led Gauss to an elegant, albeit sophisticated, formula for curvature. Intriguingly, the formula makes no reference to a surrounding space. It's intrinsic to the surface. Euclidean space has curvature zero: it's flat.

That suggested a radical idea: space can be curved without being curved *round* anything. A sphere, for instance, is clearly curved around the solid ball that it contains. To make a cylinder, you take a sheet of paper and *bend* it round in a circle, so a cylindrical surface is curved around the solid cylinder that it bounds. But Gauss did away with such outmoded thinking. He realised that you can observe the curvature of a surface without embedding it in Euclidean space.

He liked to explain this in terms of an ant living on the surface, unable to leave it, either to go inside or to launch itself into space. The surface is all the ant knows. Even light is confined to the surface, moving along geodesics, so the ant can't see that its analogue of space is curved. However, it can deduce the curvature by carrying out a survey. Tiny triangles tell it the metric of its universe, and then it can apply Gauss's formula. By crawling around measuring distances, it can *infer* that its universe is curved.

This notion of curvature differs in some respects from normal usage. For instance, a rolled-up newspaper is *not* curved, even though it looks like a cylinder. To see why, look at the letters in a headline. We see them as bent, but their shapes remain unchanged as far as their relation to the paper is concerned. Nothing got stretched, nothing got moved. The ant wouldn't notice any difference in small regions of the newspaper. As far as its metric goes, the newspaper is still *flat*. In small regions, it has the same intrinsic geometry as a plane. The angles of a small triangle, for instance, add to 180 degrees, provided you measure them within the paper. A rigid but bendy protractor is ideal.

A flat metric makes sense, once you get used to it, because that's *why* you can roll up a newspaper to make a cylinder. All lengths and angles, measured within the paper, stay the same. Locally, a newspaper-inhabiting ant can't distinguish a cylinder from a flat sheet.

The global shape – the overall form – is another matter. A cylinder has different geodesics from a plane. All geodesics of a plane are straight lines, which go on forever but never close up. On a cylinder, some geodesics can be closed, wrapping round the cylinder and returning to their starting point. Imagine using an elastic band to keep the newspaper rolled up. The band forms a closed geodesic. This kind of global difference in shape is about the overall topology – the way bits of the surface fit together. The metric just tells us about the bits.

Early civilisations were in much the same position as the ant. They couldn't go up in a balloon or an aeroplane to see the shape of the Earth. But they could make measurements and try to deduce the size and the topology. Unlike the ant, they did have some outside help: the Sun, Moon, and stars. But when it comes to the shape of the entire universe, we're in exactly the same position as the ant. We have to use analogues of the ant's geometric tricks to deduce the shape from inside.

From the ant's point of view, a surface has two dimensions. That is, it takes only two coordinates to map out any local patch. Ignoring small variations in height, terrestrial navigators need only longitude and latitude to know where they are on the Earth's surface. Gauss had a brilliant student called Bernhard Riemann, and – encouraged none too subtly by his tutor – Riemann got a bright idea: generalise Gauss's curvature formula to 'surfaces' with any number of dimensions. Since these aren't actually surfaces, he needed a new term, and he chose the German word *Mannigfaltigkeit*, which translates as 'manifold', referring to a multitude of coordinates.

Other mathematicians, notably a bunch of Italians, caught the manifold bug, creating a new subject area: differential geometry. They uncovered most of the basic ideas about manifolds. But they treated the ideas from a purely mathematical viewpoint. No one imagined that differential geometry might apply to real space.

✦

Fresh from his success with special relativity, Einstein turned his attention to the main item that was missing: gravity. He struggled for years until it dawned on him that the key is Riemannian geometry. He struggled more to master this difficult area of mathematics, helped by Marcel Grossmann, a mathematical friend who acted as guide and mentor.

Einstein realised that he needed an unorthodox variant of Riemannian geometry. Relativity allows space and time to get mixed up to some extent, even though the two concepts play different roles. In a conventional Riemannian manifold, the metric is defined using the square root of a formula that is always positive. Like Pythagoras's theorem, the metric formula is a (generalised and local) sum of squares. In special relativity, however, the analogous quantity involves *subtracting* the square of time. Einstein had to allow negative terms in the metric, leading to what we now call a pseudo-Riemannian manifold. The end result of Einstein's heroic struggles was the Einstein field equations, relating the curvature of spacetime to the distribution of matter. Matter bends spacetime; curved spacetime alters the geometry of the geodesics along which matter moves.

Newton's law of gravity doesn't describe the motion of bodies directly. It's an equation, whose solutions provide that description. Similarly, Einstein's equations don't describe the shape of the universe directly. You have to solve them. But they're nonlinear equations in ten variables, so they're hard.

We have a degree of natural intuition for Riemannian manifolds, but pseudo-Riemannian manifolds are a bit of a puzzle unless you work with them regularly. One useful simplification allows me to talk, meaningfully, about the shape of *space* – a Riemannian manifold – rather than that more slippery concept, the shape of space*time* – a pseudo-Riemannian one.

In relativity, there's no meaningful concept of simultaneity. Different observers can witness the same events happening in a different order. I see the cat leap off the windowsill just before the vase crashes to the floor; you see the vase fall before the cat leaps. Did the cat smash the vase or did the falling vase scare the cat? (We all know which one's more likely, but the cat's got a brilliant lawyer, name of Albert Einstein.)

However, although absolute simultaneity isn't possible, there's a substitute: a comoving frame. This is a fancy name for a frame of reference, or coordinate system, representing the universe as seen by a specific observer. Start where I am now, as the origin of coordinates, and travel for ten years at lightspeed to a nearby star. Define the frame so that this star is ten lightyears away from the origin and ten years into the future. Do the same for all directions and times: that's my comoving frame. We've all got one; it's just that yours can appear to be inconsistent with mine if one of us starts moving around.

If your motion looks stationary in my comoving frame, we're comoving observers. To us, the spatial shape of the universe is determined by the same fixed spatial coordinate system. The shape and size may change over time, but there's a consistent way to describe those changes. Physically, a comoving frame can be distinguished from other frames: the universe should look the same in all directions. In a frame that's not comoving, some parts of the sky are systematically redshifted, while others are blueshifted. That's why I can sensibly talk about the universe being, say, an expanding sphere. Whenever I separate out space and time in this manner, I'm referring to a comoving frame.

✦

The story now takes a bizarre swerve into the realm of mythology. Physicists and mathematicians discovered solutions of the field equations that correspond to the classical non-Euclidean geometries. Those geometries arise in spaces of constant positive curvature (elliptic), zero curvature (flat, Euclidean), and constant negative curvature (hyperbolic). So far, so good. But this correct statement rapidly transformed into a belief that these three geometries are the *only* constant-curvature solutions of the field equations.

I suspect this mistake came about because mathematicians and astronomers weren't communicating very well. The mathematical theorem states that for any fixed value of the curvature, the *metric* of constant-curvature spacetime is unique, so it was all too easy to assume that the *geometry* must also be unique. After all, doesn't the metric define the space?

No.

Gauss's ant would have made the same error had it not known the difference between a plane and a cylinder. They have the same metric, but different topologies. The metric determines only the *local* geometry, not the global. This distinction applies to general relativity, with the same implication.

A delightfully oxymoronic example is the flat torus. A torus is shaped like a doughnut with a central hole, about as far from flat as you can get. Nevertheless, there's a flat (zero curvature) manifold with doughnut topology. Start with a square, which is flat, and *conceptually* glue opposite edges together. Don't do this by physically bending the square: just identify corresponding points on opposite edges. That is, add a geometric rule to say that such points are 'the same'.

This kind of identification is common in computer games, when an alien monster rushes across one edge of the screen and reappears at the opposite edge. Programmers' jargon for this is 'wrap round': a vivid metaphor, but unwise if taken literally as an instruction. The ant would understand a flat torus perfectly: wrapping the vertical edges round turns the screen into a cylinder. Then you repeat the procedure to join the ends of the cylinder, creating a surface with the same topology as a torus. Its metric is inherited from the square, so it's flat. The natural metric on a real doughnut is different, because that surface is embedded in Euclidean space.

You can play the flat torus game with relativistic spacetime, using Minkowski's cut-down two-dimensional version of relativity. Both Minkowski's infinite plane, and a square in that plane with opposite edges identified, are flat spacetimes. But topologically, one is a plane and the other is a torus. Do the same with a cube and you get a flat 3-torus, with the same dimension as space.

Similar constructions are possible in elliptic and hyperbolic spaces. Take a chunk of the space with the right shape, glue its edges together in pairs, and you get a manifold with the same metric but different topology. Many of these manifolds are compact – they have finite size, like a sphere or a torus. Mathematicians discovered several finite spaces of constant curvature towards the end of the nineteenth century. Schwarzschild drew their work to the attention of cosmologists in 1900, explicitly citing the flat 3-torus. Aleksandr Friedmann said the same for negatively curved spaces in 1924. Unlike Euclidean and hyperbolic space, elliptic space is finite, but you can still play the same trick there to obtain spaces of constant positive curvature with different topologies. Nevertheless, for sixty years after 1930 astronomy texts repeated the myth that there are only three spaces of constant curvature, the classical non-Euclidean geometries. So astronomers worked with this limited range of spacetimes in the mistaken belief that nothing else is possible.

Cosmologists, after bigger game, turned their attention to the origin of the universe, considered only the three classical geometries of constant curvature, and found the Big Bang metric, a story we take up in the next chapter. This was such a revelation that for a long time the shape of space ceased to be a pressing issue. Everyone 'knew' it was a sphere, because that's the simplest metric for the Big Bang. However, there's little observational evidence for that shape.

Ancient civilisations thought the Earth was flat, and although they were wrong, they had some evidence: that's what it looked like. As far as the universe is concerned, we know even less than they did. But there are ideas floating around that could diminish our ignorance.

✦

If not a sphere, then what?

In 2003 NASA's Wilkinson Microwave Anisotropy Probe (WMAP) satellite was measuring a ubiquitous radio signal called the cosmic microwave background (CMB); its results are shown on page 249. A statistical analysis of fluctuations in the amount of radiation coming from different directions gives clues about how matter in the nascent universe clumped together. Before WMAP, most cosmologists thought that the universe is infinite, so it should support arbitrarily large fluctuations. But WMAP's data showed that there's a limit to the size of the fluctuations, indicative of a *finite* universe. As *Nature* said: 'You don't see breakers in your bathtub.'

The American mathematician Jeffrey Weeks analysed the statistics of these fluctuations for manifolds with a variety of topologies. One possibility fitted the data very closely, leading the media to announce that the universe is shaped like a football (US: soccer ball). This was an inevitable metaphor for a shape that goes back to Poincaré: the dodecahedral space. In the early twenty-first century footballs were made by sewing or gluing together 12 pentagons and 20 hexagons to make what mathematicians call a truncated icosahedron – an icosahedron with the corners cut off. An icosahedron is a regular solid with 20 triangular faces, arranged five to a corner. The dodecahedron, which has 12 pentagonal faces, gets into the act because the centres of the faces of an icosahedron form a dodecahedron, so both solids have the same symmetries. 'Football' is more media-friendly, albeit technically imprecise.

The surface of a football is a two-dimensional manifold. Poincaré was pioneering algebraic topology, especially in three dimensions, and discovered he'd made a mistake. To prove he'd got it wrong (unlike politicians, mathematicians do that kind of thing) he invented an analogous three-dimensional manifold. Poincaré built it by gluing two tori together, but a more elegant construction using a dodecahedron was discovered later. It's an esoteric variation on the flat 3-torus, which is made by conceptually gluing together opposite faces of a cube. Do that with a dodecahedron, but give every face a twist before gluing. The result is a three-dimensional manifold, the dodecahedral space. Like the flat 3-torus, it has no boundary: anything that falls through a face reappears through the opposite one. It's positively curved and of finite extent.

Weeks worked out the statistics of CMB fluctuations if the universe were a dodecahedral space, and found an excellent fit to the WMAP data. A group headed by Jean-Pierre Luminet deduced that a universe with that shape would have to be about 30 billion light years across – not bad. However, recent observations seem to disprove this theory, disappointing every Platonist on the planet.

It's hard to see how we could ever prove the universe to be infinite, but if it's finite, we might be able to work out its shape. A finite universe must have some closed geodesics – shortest paths that form loops, like the elastic band wrapped round a rolled newspaper. A light ray travelling along such a geodesic will eventually return to its point of origin. Point a high-powered telescope in that direction, and you'll see the back of your own head. It could take some time, mind you – as long as it takes light to go all the way round the universe – so you need to stay still and be very patient. And the head you observe might be rotated, upside down, or a mirror image of the original.

A serious mathematical analysis that takes the finite speed of light into account predicts that in such circumstances there should be repeated patterns in the CMB, in which the same fluctuations arise on distinct circles in the sky. This happens because cosmic background microwaves that reach the Earth today began their journeys from similar distances, so they originated on a sphere, the 'last scattering surface'. If the universe is finite and this sphere is larger than the universe, it wraps round and intersects itself. Spheres meet in circles, and each point on such a circle sends microwaves to Earth along two different directions, thanks to the wrap-round.

We can illustrate this effect for a two-dimensional analogue, where the geometry is simpler. If the square in the picture is big enough to contain the circle, there's no wrap-round intersection. If the square is small enough for the circle to wrap round twice, the geometry of the intersections becomes more complex.

For the flat 3-torus, the square is replaced by a cube, the circles by spheres, and the dots become circles on the faces of the cube – again identified in pairs. The dotted lines become cones. From Earth, we observe a pair of special circles in the sky, which are really the same distant circle viewed along these two directions. The fluctuations of the CMB around these two circles should be almost identical, and we can detect that using statistical correlations of the temperature fluctuations:

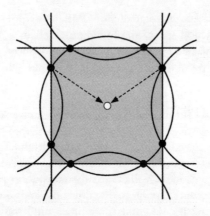

Self-intersections of the last scattering surface for a flat torus, shown
here as the large circle. Other partial circles are wrapped-round copies.
The torus is the shaded square with opposite edges identified, and Earth
is the white dot at the centre. Copies of circles meet at the black dots,
which are identified in wrap-round pairs. Dotted arrows show microwaves
arriving from the same region of space along two distinct directions.

we'd expect to see the same sequence of hot or cold patches round each
circle, where 'hot' and 'cold' mean a slightly higher or lower temperature than average.[3]

From the geometry of these circles, we can in principle deduce the
topology of the universe and distinguish the sign of the curvature:
positive, zero, or negative. So far it's not worked out in practice, though
– either because the universe isn't like that, or because it's too big for
those special circles to occur.

So what shape *is* the universe?

We haven't a clue.

16

The Cosmic Egg

In the beginning there was nothing, which exploded.
Terry Pratchett, *Lords and Ladies*

VIEWED FROM OUR COMFORTABLE, hospitable planet, teeming with life, rich in natural beauty, the rest of the universe looks hostile, remote, austere, and relatively unimportant. Viewed from distant realms of the solar system, however, our home planet shrinks to a single blue pixel[1] on a digital photograph – the famous pale blue dot, the final image snapped by *Voyager 1* in 1990. It wasn't part of the science programme, but the visionary astronomer Carl Sagan thought it would be a good idea. It's become a social and psychological icon. The probe was roughly as far away as Pluto: still in Earth's backyard, cosmically speaking. Even so, our lovely world had shrunk to an insignificant speck. From the nearest star, a camera better than anything we currently have would struggle to see our homeworld at all. From more distant stars, we might as well never have existed, for all the difference our presence would make, and the same goes for the Earth and the Sun. And when it comes to other galaxies, even our home Galaxy becomes insignificant on the cosmic scale.

It's a humbling thought, and it shows how fragile our planet really is. At the same time, it makes us wonder at the grandeur of the universe. More constructively, it also makes us wonder what else is out there, and where it all came from.

Questions such as these no doubt occurred to prehistoric humans. They certainly occurred more than 4000 years ago to civilisations such as those of China, Mesopotamia, and Egypt, which left written records.

Their answers were imaginative, if you think that attributing every-thing you don't understand to unseen deities with bizarre bodies and lifestyles counts as exercising imagination, but they were ultimately unedifying.

Over the centuries, science came up with its own theories for the origin of the universe. They were generally less exciting than world-bearing turtles, battles between the snake god and a magical cat with a sword, or gods being cut into a dozen pieces and coming back to life when reassembled. They may yet turn out to be no nearer the truth, because scientific answers are always provisional, to be discarded if new evidence contradicts them. One of the most popular theories, throughout most of the scientific era, is totally boring because nothing happens at all: the universe has no origin because it has always existed. I've always felt that this doesn't entirely dispose of the issue, because we need to explain *why* it has always existed. 'It just did' is even less satisfying than invoking a snake god. But many don't see it that way.

✦

Today most cosmologists think that the entire universe – space, time, and matter – came into existence about 13·8 billion years ago.[2] A speck of spacetime appeared from nowhere and expanded with extraordi-nary rapidity. After one billionth of a second, the initial violence tailed off enough for fundamental particles such as quarks and gluons to come into being; after another millionth of a second these particles combined to form the more familiar protons and neutrons. It took a few minutes for those to come together to make simple atomic nuclei. Atoms are nuclei plus electrons, and a further 380,000 years had to pass before electrons were thrown into the mix and atoms of the simplest elements, hydrogen, helium, and deuterium, arose. Only then could matter begin to form clumps under the influence of gravity, and eventu-ally stars, planets, and galaxies appeared. Cosmologists have calculated the timeline with exquisite precision and in considerable detail.

This scenario is the famous Big Bang, a name that Hoyle coined, somewhat sarcastically. Hoyle was a strong advocate of the main competitor at that time, the steady state theory of the universe, whose name is fairly self-explanatory. But despite the name, this wasn't a universe in which nothing happened at all. It's just that what did happen didn't cause any fundamental changes. Hoyle's view was that the universe gradually spreads out, gaining extra space as new particles quietly appear from nothing in the voids between the galaxies.

Cosmologists didn't just pluck the Big Bang from thin air. Hubble noticed a simple mathematical pattern in astronomical observations, which made it seem almost inevitable. This discovery was an unexpected by-product of his work on galactic distances, but the idea goes back to Lemaître, a few years earlier. At the start of the twentieth century, the prevailing wisdom in cosmology was very simple. Our Galaxy contained all the matter in the universe; outside it was an infinite void. The Galaxy didn't collapse under its own gravity because it rotated, so the whole arrangement was stable. When Einstein published general relativity in 1915, he quickly realised that this model of the universe was no longer stable. Gravity would cause a static universe to collapse – rotating or not. His calculation assumed a spherically symmetric universe, but intuitively the same problem would afflict any static relativistic universe.

Einstein sought a fix, and in 1917 he published it. He added an extra mathematical term to his field equations, the metric multiplied by a constant Λ (capital Greek lambda), later called the cosmological constant. This term causes the metric to expand, and by carefully adjusting the value of Λ the expansion exactly cancels gravitational collapse.

In 1927 Lemaître embarked on an ambitious project: use Einstein's equations to deduce the geometry of the entire universe. Using the same simplifying assumption that spacetime is spherically symmetric, he derived an explicit formula for this hypothetical spacetime geometry. When he interpreted the formula's meaning, Lemaître discovered that it predicted something remarkable.

The universe is expanding.

In 1927 the default view was that the universe had always existed in pretty much its current form. It just *was*, it didn't *do* anything. Just like Einstein's static universe. But now Lemaître was arguing, on the basis of a physical theory that many still thought somewhat speculative, that it *grows*. Indeed, it grows at a constant rate: its diameter increases in proportion to the passage of time. Lemaître tried to estimate the expansion rate from astronomical observations, but at that time these were too rudimentary to be convincing.

An expanding universe was a difficult notion to accept if you believed the universe to be eternal and immutable. Somehow everything there was had to be turning into increasingly more everything. Where did all the new stuff come from? It didn't make much sense. It didn't make a lot of sense even to Einstein, who, according to Lemaître, said something to the effect that 'Your calculations are right, but your physics is abominable.' It may not have helped that Lemaître called his theory 'the Cosmic Egg exploding at the moment of the creation', especially since he was a Jesuit priest. It all seemed a bit biblical. However, Einstein wasn't totally dismissive, and suggested that Lemaître should consider more general expanding spacetimes without the strong assumption of spherical symmetry.

✦

Within a few years, evidence emerged to vindicate Lemaître. In Chapter 11 we saw how Hubble's 'computer' Leavitt, cataloguing the brightness of thousands of stars, noticed a mathematical pattern in one particular type of star called a Cepheid variable. Namely, the intrinsic brightness, or luminosity, is related, in a specific mathematical way, to the period over which the brightness repeats. This lets astronomers use Cepheids as standard candles, whose apparent brightness can be compared to their actual brightness, telling us how far away they must be.

At first the method was limited to stars in our own Galaxy, because telescopes were unable to resolve individual stars in other galaxies, let alone observe their spectra to see whether they might be Cepheids. But as telescopes improved, Hubble set his sights on a big question: how far away are galaxies? As described in Chapter 12, in 1924 he used Leavitt's distance–luminosity relation to estimate the distance to the Andromeda Galaxy M31. His answer was 1 million light years; the current estimate is 2·5 million.

Leavitt had taken a small step for a woman, but a giant leap up the cosmic distance ladder. Understanding variable stars linked the geometric method of parallax to observations of apparent brightness. Now Hubble was to make a further leap, opening up the prospect of mapping any cosmological distance, however great.

This possibility stemmed from an unexpected discovery by Vesto Slipher and Milton Humason: the spectra of many galaxies are shifted towards the red end of the spectrum. It seemed likely that the shift is caused by the Doppler effect, so galaxies must be moving away from us. Hubble took 46 galaxies known to contain Cepheids, making it possible to infer their distances, and plotted the results against the amount of redshift. What he found was a straight line, indicating that a galaxy recedes with a speed proportional to its distance. In 1929 he stated this relationship as a formula, now called Hubble's law. The constant of proportionality, Hubble's constant, is about 70 km/s per megaparsec. Hubble's first estimate was seven times as big.

In fact, the Swedish astronomer Knut Lundmark had the same idea in 1924, five years before Hubble. He used the apparent sizes of galaxies to infer how far away they are, and his figure for the 'Hubble' constant was within 1% of today's – far better than Hubble's. However, his work was ignored because his methods hadn't been cross-checked using independent measurements.

Now astronomers could estimate the distance of any object from its spectrum, provided they could spot enough spectral lines to deduce the redshift. Virtually all galaxies exhibit redshift, so we can calculate how far away they are. All of them are moving away from us. So either the Earth lies at the centre of a huge expanding region, violating the Copernican principle that we're not special, or the entire universe is getting bigger, and aliens in another galaxy would observe the same behaviour.

Hubble's discovery was evidence for Lemaître's cosmic egg. If you run an expanding universe backwards in time, it all condenses to a point. Restoring time to its usual direction, the universe must have started as a point. The universe doesn't emerge from an egg: it *is* an egg. The egg appears from nowhere, and grows. Both space *and time* spring into existence from nothing, and once they exist, today's universe evolves.

When Hubble's observations convinced Einstein that Lemaître had been right all along, he realised that he could have *predicted* cosmic expansion. His static solution could have been modified into an expanding one, and the expansion would prevent gravitational collapse. That pesky cosmological constant Λ was unnecessary: its role had been to prop up an incorrect theory. He removed Λ from his theory and later said that including it had been his biggest blunder.

The upshot of all this work is a standard model of the spacetime geometry of the universe, the Friedmann–Lemaître–Robertson–Walker metric, put together in the 1930s. It's actually a family of solutions, each giving a possible geometry. It includes a parameter specifying the curvature, which can be zero, positive, or negative. Every universe in the family is homogeneous (the same at every point) and isotropic (the same in every direction) – the main conditions assumed to derive the formula. Spacetime can be expanding or contracting, and its underlying topology can be simple or complex. The metric also includes an optional cosmological constant.

✦

Because time comes into existence with the Big Bang, there's no logical need to say what occurred 'before'. There *was* no before. Physics was ready for this radical theory, because quantum mechanics shows that particles can arise spontaneously from nothing. If a particle can do it, why not a universe? If space can do it, why not time? Cosmologists now believe this is basically correct, although they're starting to wonder whether 'before' can be dismissed that easily. Detailed physical calculations allow the construction of a complex, very precise timeline, according to which the universe came into existence 13·8 billion years ago as a single point, and has been expanding ever since.

One intriguing feature of the Big Bang is that individual galaxies, even gravitationally bound clusters of them, are *not* expanding. We can estimate the sizes of distant galaxies, and the statistical distribution of sizes is much the same as for nearby galaxies. What's happening is far weirder. The distance scale of *space* is changing. Galaxies move apart because more space appears between them, not because they're travelling in opposite directions through a fixed amount of space.

This leads to some paradoxical effects. Galaxies more than 14·7 billion light years away are moving so fast that, relative to us, they're travelling faster than light. Yet we can still see them.

There seem to be three things wrong with these claims. Since the universe is only 13·8 billion years old, and originally it was all in the same location, how can anything be 14·7 billion light years away? It would have to move faster than light, which relativity forbids. For the same reason, galaxies can't now be exceeding the speed of light. Finally, if they were doing that, we wouldn't see them.

To understand why the claims make sense, we have to understand a bit more about relativity. Although it forbids matter moving faster than light, that limit is with respect to the surrounding space. However, relativity does not forbid *space* moving faster than light. So a region of space could exceed lightspeed, while matter within it would remain below lightspeed relative to the space that contains it.[3] In fact, the matter could be at rest relative to its surrounding space while the space was belting along at ten times lightspeed. Just as we sit in peace and comfort, drinking coffee and reading a newspaper, inside a passenger jet doing 700 kph.

That's also how those galaxies can be 14·7 billion light years away. They didn't travel that distance as such. The amount of space between us and them has grown.

Finally, the light by which we observe these distant galaxies isn't the light they are currently emitting.[4] It's light they emitted in the past, when they were closer. That's why the observable universe is bigger than we might expect.

You might wish to get a coffee and a newspaper while you think about that.

Here's another curious consequence.

According to Hubble's law, distant galaxies have greater redshifts, so they must be moving faster. At first sight this is inconsistent with the Friedmann–Lemaître–Robertson–Walker metric, which predicts that the expansion rate should slow down as time passes. But again we must think relativistically. The further away a galaxy is, the longer its light has taken to reach us. Its redshift *now* indicates its velocity *then*. So Hubble's law implies that the further into the past we look, the faster space was expanding. In other words, the expansion was initially rapid, but then slowed down in accordance with Friedmann–Lemaître–Robertson–Walker.

That makes perfect sense if all of the expansion was imparted in the initial Bang. As the universe began to grow, its own gravity started to pull it back again. Observations indicate that until about 5 billion years ago, that's what happened. The calculations are based on Hubble's law, which tells us that the expansion rate grows by 218 kilometres per second for every extra million light years of distance. That is, it increases by 218 kilometres per second every million years into the past, so equivalently it has decreased by 218 kilometres per second for every million years after the Big Bang.

We'll see in Chapter 17 that this slowdown in the expansion seems to have been reversed, so it's speeding up again, but let's not go into that right now.

✦

The next step was to obtain independent evidence confirming the Big Bang. In 1948 Ralph Alpher and Robert Herman predicted that the Big Bang ought to have left an imprint on the radiation levels of the universe, in the form of a uniform cosmic microwave background radiation. According to their calculations, the temperature of the CMB – that is, the temperature of a source that could produce that level of radiation – is about 5 K. In the 1960s Yakov Zel'dovich and Robert Dicke independently rediscovered the same result. Astrophysicists A.G. Doroshkevich and Igor Novikov realised in 1964 that in principle the CMB can be observed and used to test the Big Bang.

In the same year, Dicke's colleagues David Wilkinson and Peter Roll started to build a Dicke radiometer to measure the CMB. This is a radio receiver that can measure the average power of a signal in some range of frequencies. But before they could complete the work, another team beat them to it. In 1965 Arno Penzias and Robert Wilson used a Dicke radiometer to construct one of the first radio telescopes. Investigating a persistent source of 'noise', they realised that its origin was cosmological, not a fault in their electronics. The noise had no specific location; instead, it was evenly distributed over the entire sky. Its temperature was roughly 4·2 K. This was the first observation of the CMB.

The explanation of the CMB was hotly debated in the 1960s, and physicists who favoured the steady state theory suggested it was starlight scattered from distant galaxies. But by 1970 the CMB was widely accepted as evidence for the Big Bang. Hawking called this observation 'the final nail in the coffin of the steady state theory'. The clincher was its spectrum, which looked just like black-body radiation, contrary to the steady state theory. The CMB is now thought to be a relic of the universe when it was 379,000 years old. At that time its temperature dropped to 3000 K, making it possible for electrons to combine with protons to form hydrogen atoms. The universe became transparent to electromagnetic radiation. Let there be light!

Theory predicts that the CMB should not be exactly uniform in all directions. There ought to be very small fluctuations, estimated to be of the order of 0·001–0·01%. In 1992 NASA's Cosmic Background Explorer (COBE) mission measured these inhomogeneities. Their detailed structure was further revealed by NASA's WMAP probe. These details have become the main way to compare reality with predictions from various versions of the Big Bang and other cosmological scenarios.

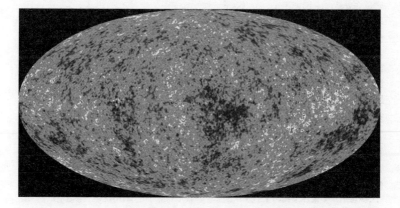

Cosmic microwave background radiation measured by WMAP.
The map shows temperature fluctuations dating from shortly after
the Big Bang, seeds of irregularity that grew to create galaxies.
Temperatures differ from the average by a mere 200 millionths K.

✦

When my family was travelling in France some years ago, we were amused to see a sign for the *Restaurant Univers*. Unlike Douglas Adams's fictional creation *The Restaurant at the End of the Universe*, poised forever at the very brink of the ultimate point in space and time, this was a perfectly normal eatery attached to the Hotel Univers. Which, in turn, was a perfectly normal hotel in Reims, at just the right point in space and time for four tired and hungry travellers.

The scientific issue that motivated Adams's fictional restaurant was: how will the universe end? Not with a rock concert of cosmological proportions, which was his answer. That might be a fitting end for humanity, but perhaps not one we should inflict on any other civilisations that might be out there.

Perhaps it won't end at all. It could keep on expanding forever. But if it does, everything will slowly run down, galaxies will recede from each other until light can't pass between them, and we will be alone in the cold and the dark. However, according to Freeman Dyson, complex 'life' could still continue to exist, despite this so-called 'heat death' of the universe. But it would be very *slow* life.

Less disappointingly for science fiction fans, the universe could collapse in a reverse Big Bang. It might even collapse back to a point. Or its end could be messier, a Big Crunch in which matter gets shredded as dark energy rips the fabric of spacetime apart.

That could be The End. But it's conceivable that after collapsing, the universe could bounce back into existence. This is the theory of the oscillating universe. James Blish used it at the end of *A Clash of Cymbals*. Maybe the fundamental constants of physics would be different after the bounce; some physicists think so. Others don't. Maybe our universe will bud off babies, just like their mother or totally different. Maybe it won't.

Mathematics lets us explore all of these possibilities, and may one day help to decide between them. Until then, we can only speculate on the end of all things – or not, as the case may be.

The Big Blow-Up

> If I had been present at the creation, I would have given some
> useful hints for the better arrangement of the Universe.
>
> Alfonso the Wise, king of Castile (*attributed*)

A FEW YEARS ago, the Big Bang theory of the origin of the universe
fitted all the important observations. In particular, it predicted the
temperature of the cosmic microwave background radiation, an early
success that contributed significantly to the acceptance of the theory.[1]
On the other hand, observations were few and far between. As astronomers obtained more detailed measurements, and made more extensive
calculations to see what the Big Bang predicted, discrepancies began
to emerge.

We saw in Chapter 15 that today's universe has a lot of large-
scale structure, with vast filaments and sheets of galaxies surrounding
even vaster voids, much like the foam in a glass of beer, with galaxies
occupying the surfaces of the bubbles, and voids corresponding to the
air inside them. Calculations indicate that the current estimate for the
age of the universe, 13·8 billion years, doesn't allow enough time for
matter to have become as clumpy as it is today. It's also too short a time
to explain the current flatness of space. Fixing both is tricky because
the flatter you make space, the less clumpy matter is likely to become,
and clumpier matter makes space more curved.

The prevailing cosmological wisdom is to postulate an even Bigger
Bang, known as inflation. At a critical juncture, very early in its
existence, the nascent universe expanded to a huge size in an extraordinarily short time.

Other deficiencies of the original Big Bang theory led cosmologists to bolt on two further assumptions: dark matter, a form of matter utterly different from normal matter, and dark energy, a form of energy that causes the expansion of the universe to speed up. In this chapter I'll consider inflation and dark energy. I'll save dark matter for the next chapter, because there's a lot to say about it.

Cosmologists are very confident about the current theory, known as the ΛCDM (Lambda cold dark matter) model, or standard model of cosmology. (Recall that Λ is the symbol for Einstein's cosmological constant.) They're confident because the combination of the classical Big Bang, inflation, dark matter, and dark energy agrees with most observations, in considerable detail. However, there are significant problems with all three add-ons, and some rethinking may be in order.

In this chapter and the next I'm going to tell the conventional stories first, sketching the observations that motivated the three bolt-on extras, and describing how they explain those observations. Then I'm going to take a critical look at the resulting standard model of cosmology, outlining some of the problems that still remain. Finally, I'm going to describe some alternatives to the standard model that have been proposed, and see how well they fare in comparison.

✦

Chapter 16 described the main evidence for the Big Bang, add-ons included: the structure of the cosmic microwave background (CMB). The most recent measurements from WMAP show that the CMB is *almost* uniform, departing from the average by no more than 200 millionths of a degree kelvin. Small fluctuations are what the Big Bang predicts, but these are *too* small – so small that the current clumpiness of the universe has not had sufficient time to evolve. This statement is based on computer simulations of mathematical models of the evolution of the universe, mentioned in Chapter 15.

One way to fix the problem is to modify the theory so that the early universe is clumpier to begin with. But this idea runs into a second difficulty, almost the opposite to the first. Although, today, *matter* is too clumpy to fit the standard Big Bang, *spacetime* isn't clumpy enough. It's nearly flat.

Cosmologists were also worried by a deeper issue, the horizon problem, which Misner pointed out in the 1960s. The standard Big Bang predicts that parts of the universe that are too far apart to have a causal effect on each other should nevertheless have a similar distribution of matter and a similar CMB temperature. Moreover, this should become apparent to an observer because the cosmological horizon – how far they can see – increases as time passes. So regions that used not to be causally connected will become so later on. The problem then is: how can these regions 'know' what distribution and temperature they ought to have? So it's not just that spacetime is too flat: it's also *uniformly* flat over regions that are too big to have communicated with each other.

In 1979 Alan Guth came up with a cunning idea that sorts out both issues. It makes spacetime flat while allowing matter to stay clumpy, and it solves the horizon problem. To describe it, we need to know about vacuum energy.

In today's physics, a vacuum isn't just empty space. It's a seething cauldron of virtual quantum particles, appearing from nothing in pairs and then annihilating each other again before anyone can observe them. This is possible in quantum mechanics because of the Heisenberg uncertainty principle, which says that you can't observe a particle's energy at a specific time. Either the energy, or the interval of time, has to be fuzzy. If energy is fuzzy, it need not be conserved at every instant. Particles can borrow energy, and then pay it back, during that brief timespan. If time is fuzzy, its absence isn't noticed.

This process – or something else, physicists aren't certain – creates a bubbling field of background energy, everywhere in the universe. It's small, about a billionth of a joule per cubic metre. Enough to power one bar of an electric fire for one trillionth of a second.

Inflation proposes that widely separated regions of spacetime have the same matter distribution and temperature because in the past they *were* able to communicate with each other. Suppose that now-distant regions of the universe were once close enough to interact. Suppose also that at that time the vacuum energy was bigger than it is now. In such a state, the observable horizon doesn't increase; instead, it remains constant. If the universe then undergoes a rapid expansion, nearby observers quickly become separated, and everything becomes homogeneous. Essentially, any localised bumps and dips that existed before inflation set in are suddenly spread out over a truly gigantic volume of spacetime. It's like placing a lump of butter on a small piece of toast, and then making the toast suddenly grow to an enormous size. As the butter spreads with it, you get a thin, almost uniform layer.

Don't try this at home.

✦

Both the very early onset and the very rapid expansion are needed to make the inflationary sums come out right. So what causes this rapid growth – an explosion far more impressive than the wimpy Big Bang that started it all? The answer is, an inflaton field. That's not a typo: an inflaton is a hypothetical particle. In quantum theory, fields and particles go hand in hand. A particle is a localised lump of field, and a field is a continuous sea of particles.

Guth wondered what would happen if space is uniformly filled by an unnoticed quantum field – the hypothetical inflaton field. His calculations showed that such a field generates negative pressure – that is, it provides an outward push. Brian Greene suggests the analogy of carbon dioxide gas in a bottle of champagne. Pop the cork, and the gas expands with great rapidity, creating those desirable bubbles. Pop the cork on the universe, and the inflaton field expands even more rapidly. The new twist is that you don't need a cork: instead, the entire bottle (the universe) can expand, very quickly, by a gigantic amount. The current theory is that between 10^{-36} and 10^{-32} seconds after the Big Bang, the volume of the universe multiplied by a factor of at least 10^{78}.

The good news is that the inflationary scenario – more precisely, some of the numerous variations on the original idea that have since been proposed – fits many observations. That's not totally surprising, because it was designed to fit some key observations, but, reassuringly, it fits a lot of others as well. Job done, then? Well, maybe not, because the bad news is that no one has ever detected either an inflaton, or any trace of the field it allegedly supports. It's a quantum rabbit not yet extracted from a cosmological hat, but it would be a very attractive rabbit if it could be persuaded to poke its waffly nose above the brim.

However, in the last few years it's started to look a lot less attractive. As physicists and cosmologists asked deeper questions about inflation, problems emerged. One of the biggest is eternal inflation, discovered by Alexander Vilenkin. The usual explanation of the structure of our universe assumes that the inflaton field switches on once, very early in the evolution of the universe, and then *stays* switched off. However, if an inflaton field exists at all, it can swing into action anywhere and at any time. This tendency is known as eternal inflation. It implies that our region of the universe is just one inflated bubble in a bubble bath of cosmic foam, and a new period of inflation might start in your living room this afternoon, instantaneously blowing up your television and the cat[2] by a factor of 10^{78}.

There are ways to fix this using variations on Guth's original idea, but they require extraordinarily special initial conditions for our universe. Just how special can be inferred from another curious fact: there exist other special initial conditions that lead to a universe just like ours *without* inflation occurring. Both types of condition are rare, but not equally so. Roger Penrose[3] showed that that the initial conditions that produce our universe without invoking inflation outnumber those that produce inflation by a factor of one googolplex – $10^{10^{100}}$. So an explanation of the current state of the universe that *doesn't* invoke inflation is overwhelmingly more plausible than one that does. Penrose used a thermodynamic approach, and I'm not sure that's appropriate in this context, but Gary Gibbons and Neil Turok used a different method: reverse time and unwind the universe back to its initial state. Again, almost all such states don't involve inflation.

Most cosmologists remain convinced that the theory of inflation is essentially correct because its predictions agree remarkably well with observations. It would be premature to discard it because of the difficulties I've mentioned. However, those difficulties strongly suggest that the current concept of inflation has serious deficiencies. It may be pointing us in the right direction, but it's by no means the final answer.

✦

There are two other problems with the standard model of the origin of the universe. One, set up in Chapter 12, is: the outer regions of galaxies are rotating much too fast to hold together if Newtonian (or, it's generally believed, Einsteinian) gravity applies. The standard answer to this is dark matter, treated in detail in the next chapter.

The other is how the expansion rate of the universe changes over time. Cosmologists had expected it either to remain constant, leading to an 'open' universe that never stops growing, or to slow down as gravity hauls the expanding galaxies back together again to form a 'closed' universe. But in 1998 the High-z Supernova Search Team's observations of redshift in Type Ia supernovas revealed that the expansion is *speeding up*. Their work won the 2011 Nobel Prize in Physics, and the actual result (unlike inflation and dark matter) is not particularly controversial. What is controversial is its explanation.

Cosmologists attribute the accelerating expansion of the universe to a presumed energy source, which they call 'dark energy'. One possibility is Einstein's cosmological constant Λ. A positive value for Λ, inserted into the equations, creates the observed rate of acceleration. If this is correct, Einstein's biggest blunder wasn't putting the cosmological constant into his field equations, it was taking it out again. To fit observations, its value must be extremely small: about 10^{-29} grams per cubic centimetre when energy is expressed as a mass via Einstein's famous equation $E = mc^2$.

A possible physical reason why Λ should be greater than zero comes from quantum mechanics: vacuum energy. Recall that this is a natural repulsive effect created by virtual particle/antiparticle pairs blinking into existence and then annihilating each other so rapidly that the particles themselves can't be detected. The only problem is that according to today's quantum mechanics, the vacuum energy should be 10^{120} times bigger than the value of Λ that fits the rate of acceleration.

The South African mathematician George Ellis has pointed out that the presence of dark energy is deduced from the observations by assuming that the universe is correctly described by the standard Friedmann–Lemaître–Robertson–Walker metric, in which Λ may (by a change of coordinates) be interpreted as dark energy. We saw that this metric is derived from two simple requirements: the universe should be homogeneous and isotropic. Ellis has shown that a lack of homogeneity can explain the observations without assuming dark energy exists.[4] The universe is inhomogeneous on the scale of its voids and clusters, which are much larger than galaxies. On the other hand, the standard cosmological model assumes that on extremely large scales, these inhomogeneities are smoothed out, much as a foam looks smooth if you don't look closely enough to see the bubbles. Cosmologists therefore compare the High-z observations with the predictions of this smoothed model.

A delicate mathematical issue now surfaces, which seems to have been neglected until recently: is an exact solution of the smoothed model close to a smoothed solution of the exact model? The former corresponds to the prevailing theory, the latter to how we compare it to observations. The tacit assumption is that these two mathematical processes produce pretty much the same result – a version of the modelling assumption, common in mathematical physics and applied mathematics, that small terms in equations can be neglected without having much effect on the solutions.

This assumption is often correct, but not always, and indications are that here it can produce incorrect results. Thomas Buchert[5] has shown that when Einstein's equations for the clumpy small-scale structure are averaged to derive a smooth large-scale equation, the result is not the same as the Einstein equations for the smooth large-scale model. Instead, it has an extra term, a repulsive 'back reaction' creating an effect that mimics dark energy.

Observations of distant cosmological sources can also be misinterpreted, because gravitational lensing can focus light and make it seem brighter than it should be. The average effect of such focusing, over all distant objects, is the same for detailed small-scale clumpy models and their large-scale averages, which at first sight is encouraging. However, the same is not true for individual objects, and those are what we observe. Here, the correct mathematical procedure is to average over light paths, not over ordinary space. Failing to do so can change the apparent luminosity, but exactly how depends sensitively on the distribution of matter. We don't know this accurately enough to be sure what happens. But it seems that the evidence for the acceleration of the expansion of the universe may be unreliable for two distinct but related reasons: the usual smoothing assumptions can give incorrect results, both for theory and for observations.

Another way to explain the High-z observations, without invoking dark energy, is to tinker with Einstein's field equations. In 2009 Joel Smoller and Blake Temple used the mathematics of shockwaves to show that a slightly modified version of the field equations has a solution in which the metric expands at an increasing rate.[6] This would account for the observed acceleration of galaxies without invoking dark energy.

In 2011, in a special issue on general relativity of a Royal Society journal, Robert Caldwell[7] wrote: 'To date, it appears entirely reasonable that the [High-z] observations may be explained by new laws of gravitation.' Ruth Durrer[8]described the evidence for dark energy as weak: 'Our single indication for the existence of dark energy comes from distance measurements and their relation to redshift.' In her view the rest of the evidence establishes only that distance estimates from redshift are larger than expected from the standard cosmological model. The observed effect might not be an acceleration, and even if it is, there's no compelling reason to suppose the cause is dark energy.

✦

Although mainstream cosmology continues to focus on the standard model – Big Bang as described by the ΛCDM metric, plus inflation, dark matter, and dark energy – mutterings of discontent have been growing for some time. At a conference on alternatives in 2005, Eric Lerner said: 'Big Bang predictions are consistently wrong and are being fixed after the event.' Riccardo Scarpa echoed this view: 'Every time the basic Big Bang model has failed to predict what we see, the solution has been to bolt on something new.'[9] Both were signatories to an open letter a year earlier, warning that research into alternative cosmological theories wasn't being funded, suppressing scientific debate.

Such complaints might be merely sour grapes, but they were based on some disconcerting evidence – not just philosophical objections to the three add-ons. The Spitzer space telescope has spotted galaxies with such a high redshift that they date back to less than a billion years after the Big Bang. As such, they ought to be dominated by young, superhot blue stars. Instead, they contain too many old, cool red stars. This suggests that these galaxies are older than the Big Bang predicts, and therefore so is the universe. In support, some stars today appear to be older than the universe. They're red giants, so large that the time required to burn enough hydrogen to reach that state should be much longer than 13·8 billion years. Moreover, there are huge superclusters of galaxies at high redshift, which wouldn't have had time to organise themselves into such large structures. These interpretations are disputed, but the third is especially hard to explain away.

If the universe is much older than currently thought, how can we explain the observations that led to the Big Bang theory? The main ones are redshift and the CMB, along with a lot of finer detail. Perhaps the CMB isn't a relic of the origin of the universe; just starlight that has bounced around the universe for aeons, being absorbed and radiated again. General relativity concentrates on gravity, whereas this process involves electromagnetic fields as well. Since most of the matter in the universe is plasma, whose dynamics is driven by electromagnetism, it seems strange to ignore these effects. However, plasma cosmology lost support in 1992 when COBE data showed that the CMB has a black-body spectrum.[10]

What of redshift? It certainly exists, and is pretty much ubiquitous, and it varies with distance. In 1929 Fritz Zwicky suggested that light loses energy as it travels, so the greater the distance it traverses, the greater the redshift. This 'tired light' theory is said to be incompatible with time dilation effects that fit a cosmological (expansion) origin of redshift, but similar theories with different mechanisms avoid this particular problem.

Gravity reduces the energy of photons, which shifts spectra towards the red end. The gravitational redshift caused by ordinary stars is very small, but black holes, such as those at the centre of galaxies, have a bigger effect. In fact, large-scale fluctuations in the CMB (as measured by WMAP) are mainly caused by gravitational redshift. However, the effect is still too small. Nevertheless, Halton Arp has argued for years that redshift could result from the effect of strong gravity on light – a theory that has conventionally been dismissed without any satisfactory disproof. This alternative even predicts the correct temperature for the CMB. And it avoids assuming that space expands but galaxies don't – even though they're mainly empty space.[11]

Alternatives to the Big Bang continue to flood in. One of the latest, proposed in 2014 by Saurya Das and developed in conjunction with Ahmed Ali,[12] is based on David Bohm's reformulation of quantum mechanics, which removes the element of chance. Bohmian quantum theory is unorthodox but fairly respectable; those who dismiss it do so because it's equivalent in most respects to the standard approach, differing mainly in interpretation, rather than provably wrong. Ali and Das dispute the usual argument in favour of the Big Bang, which runs the expansion of the universe backwards to produce an initial singularity. They point out that general relativity breaks down before the singularity is reached, but cosmologists continue to apply it as though it remains valid. Instead, Ali and Das employ Bohmian quantum mechanics, in which the trajectory of a particle makes sense and can be calculated. This leads to a small correction term in Einstein's field equations, which removes the singularity. In fact, the universe could always have existed without any conflict with current observations.

Competitors to the Big Bang have to pass some stringent tests. If the universe has been around forever, most of its deuterium should have disappeared through nuclear fusion, but it hasn't. On the other hand, if the lifetime is finite but there was no Big Bang, there wouldn't be enough helium. These objections rest on specific assumptions about the far past of the universe, however, and ignore the possibility that something just as radical as the Big Bang – but different – might have occurred. No really strong case for a specific alternative explanation has yet emerged, but the Big Bang doesn't look too solid either. I suspect that fifty years from now, cosmologists will be promoting entirely different theories of the origin of the universe.

✦

The prevailing public view of cosmology, in which the origin of the universe has been solved once and for all by the Big Bang, fails to reflect deep divisions among experts, and ignores the exciting but confusing variety of alternatives that are being contemplated and argued about. There's also a tendency to overstate the implications of the latest idea or discovery – orthodox or not – before anyone else has had time to think about it critically. I've lost count of the number of times that some group of cosmologists has announced a definitive proof that inflation exists, only to be contradicted a few weeks or months later by a different interpretation of the data or the revelation of an error. The same can be said even more strongly of dark matter. Dark energy seems more robust, but even that's debatable.

A topical example of confirmation quickly becoming retraction was the announcement in March 2014 that the BICEP2 experiment had observed patterns in light from distant sources, relics of the Big Bang, that proved beyond any shadow of doubt that the inflationary theory of the universe is correct. And, as a bonus, also confirmed the existence of gravitational waves, predicted by relativity but never before observed. BICEP stands for 'Background Imaging of Cosmic Extragalactic Polarization', and BICEP2 is a special telescope that measures the CMB. At the time the announcement was greeted with great excitement; either of those discoveries would be a shoo-in for a Nobel Prize. But almost immediately other groups started to wonder whether the true cause of the patterns was interstellar dust. This wasn't just carping: they'd been thinking about that issue for some time.

By January 2015 it became clear that at least half of the signal that BICEP2 had detected was caused by dust, not inflation. The team's claims have now been withdrawn altogether, because once the contribution known to come from dust is excluded, what remains of the signal is no longer statistically significant. Turok, an early critic of the BICEP2 results, also pointed out that far from confirming inflation, the corrected data *disprove* several of the simpler inflationary models.

This story is embarrassing for the BICEP2 team, which has been criticised for premature claims. Jan Conrad, writing in *Nature*,[13] remarked that the scientific community must 'ensure that enticing reports of false discoveries do not overwhelm more sober accounts of genuine scientific breakthroughs'. On the other hand, these events show real science in action, warts and all. If no one is allowed to get things wrong, no progress will ever be made. It also illustrates scientists' willingness to *change their minds* when new evidence comes along or old evidence is shown to be misleading. The BICEP2 data are good science; only the interpretation was wrong. In today's world of instant communications it's impossible to sit on what looks like a big discovery until it's been verified to the hilt.

Nonetheless, cosmologists routinely make breathtaking claims on the basis of little genuine evidence, and display supreme confidence in ideas that have only the shakiest foundations. Hubris begets Nemesis, and Nemesis is hanging around a lot in the wings nowadays. The spirit of divine retribution may yet take centre stage.

The Dark Side

There was nothing in the dark that wasn't there when the lights were on.

Rod Serling, *The Twilight Zone*, Episode 81: 'Nothing in the Dark'

CHAPTER 12 ENDED with the word 'oops'. It was a comment on the discovery that the rotation speeds of galaxies don't make sense. Near the centre, a galaxy spins quite slowly, but the speed at which stars rotate increases the further out you get, and then levels off. However, both Newtonian and Einsteinian gravity require the rate of spin to decrease in the outer reaches of the galaxy.

Cosmologists solve this riddle by positing that most galaxies sit in the middle of a vast spherical halo of invisible matter. At one time they hoped it was just ordinary matter that didn't emit enough light for us to see it from intergalactic distances, and called it cold dark matter. Maybe it was just a lot of gas, or dust, shining too faintly for us to notice. But as more evidence came in, this easy way out ceased to be tenable. Dark matter, as currently conceived, is unlike anything we've ever encountered, even in high-energy particle accelerators. It's mysterious, and there must be an awful lot of it.

Recall that in relativity, mass is equivalent to energy. The standard model of cosmology, plus data from the European Space Agency's Planck cosmology probe, suggest that the total mass/energy of the known universe comprises a mere 4·9% normal matter but 26·8% of dark matter. That leaves an even bigger 68·3%, attributed to dark energy. There seems to be five times as much dark matter as normal matter, and in galactic-scale regions of the universe the mass of dark matter plus the effective mass of dark energy is *twenty times* the mass of normal matter.

The argument in favour of dark matter, in huge amounts, is simple and straightforward. Its existence is inferred by comparing the predictions of the Kepler equation with observations. This formula occupied centre stage in Chapter 12. It states that the total mass of a galaxy, out to a given radius, is equal to that radius multiplied by the square of the rotational velocity of the stars at that distance, and divided by the gravitational constant. The pictures on page 185 show that observations are in serious disagreement with this prediction. Near the galactic core the observed rotation speed is too small; further out, it's too big. In fact, rotation curves remain roughly constant out to much greater distances than the observable matter, which is basically what we can see through the light it emits.

If you use observed speeds to calculate masses, you find that an awful lot of mass must lie beyond the visible radius. To rescue the Kepler equation, whose derivation seemed foolproof, astronomers were forced to postulate the existence of large quantities of unobserved dark matter. They've stuck to that story ever since.

The anomalous behaviour of galactic rotation curves was the first evidence, and is still the most convincing, that very large amounts of invisible matter must exist in the universe. Further observations and other gravitational anomalies lend weight to the idea, and indicate that dark matter is not just ordinary matter that happens not to emit light. It has to be an entirely different kind of matter, which mainly interacts with everything else through the force of gravity. Therefore it must be made from subatomic particles that are completely different from anything ever observed in particle accelerators.

Dark matter is a type of matter unknown to physics.

✦

It's reasonable that a lot of the matter in the universe might not be observable, but the dark matter story currently lacks a punchline. The real clincher would be to create new particles with the required properties in an accelerator, such as the Large Hadron Collider (LHC). This impressive apparatus recently made the epic observation of the Higgs boson, a particle that explains why many (but not all) particles have mass. But to date, no dark matter particles have been detected in accelerator experiments. Neither has anything been found in cosmic rays – high-energy particles from outer space that hit the Earth in huge quantities.

So the universe is full of this stuff, it's far more common than ordinary matter – but everywhere we look, we see only ordinary matter.

Physicists point to precedents. Weird hypothetical particles have a good track record. The classic case is the neutrino, whose existence was inferred by applying the law of conservation of energy to certain particle interactions. It had to be very strange compared to the then known particles: no electric charge, hardly any mass, able to pass almost unhindered through the entire body of the Earth. It sounded ridiculous, but experiments have detected neutrinos. Some scientists are now taking the first steps towards neutrino astronomy, using these particles to probe the distant realms of the universe.

On the other hand, plenty of hypothetical particles have turned out to be figments of theorists' overstretched imaginations.

For a time it was thought that maybe we were failing to spot a lot of perfectly ordinary 'cold dark matter' – massive compact halo objects, aka MACHOs. This term covers any kind of body that is made of normal matter, emits very little radiation, and could exist in a galactic halo, such as brown dwarfs, faint red and white dwarfs, neutron stars, black holes … even planets. When the enigma of rotation curves first became apparent, this kind of matter was the obvious candidate for an explanation. However, MACHOs seem insufficient to account for the vast amount of unobserved matter that cosmologists believe must be present.

A totally new type of particle is needed. It has to be something that theorists have thought about, or can start to think about, and by definition it has to be something we don't yet know exists. So we're thrown headlong into the realm of speculation.

One possibility is a range of hypothetical particles known as weakly interacting massive particles (WIMPs). The proposal is that these particles emerged from the dense superheated plasma of the early universe, and interact with ordinary matter only via the weak nuclear force. Such a particle fits the bill if it has an energy of about 100 GeV. The theory of supersymmetry, one of the leading candidates for the unification of relativity and quantum mechanics, predicts a new particle with exactly those properties. This coincidence is known as the WIMP miracle. When the LHC began its observations, theorists hope it would see a whole raft of new supersymmetric partners of the known particles.

Zilch.

The LHC has explored a range of energies that includes 100 GeV, and seen absolutely nothing unaccounted for by the standard model.

Several other WIMP-hunting experiments have also drawn a blank. Not a trace has been detected in emissions from nearby galaxies, and WIMPs are notably absent in laboratory experiments aimed at spotting remnants of their collisions with nuclei. The Italian DAMA/LIBRA detector keeps seeing what look like signals of WIMPs, which should generate a burst of light when they hit a sodium iodide crystal. These signals occur as regular as clockwork every June, suggesting that the Earth is passing through a bunch of WIMPs at some specific position in its orbit. The problem is, other experiments ought to be detecting these WIMPs too – and they don't. DAMA is seeing something, but probably not WIMPs.

Might dark matter be a far heavier kind of particle, a WIMPZILLA? Perhaps. The BICEP2 radio telescope provides convincing evidence that the early universe had enough energy to create the elusive inflaton, and this could have decayed into WIMPZILLAs. All very well, but these beasts are so energetic that we can't make them, and they pass through ordinary matter as if it's not there, so we can't observe them. But maybe we can spot what they produce when they hit other stuff: the IceCube experiment at the North Pole is looking. Of the 137 high-energy neutrinos it had detected in mid-2015, three might have been generated by WIMPZILLAs.

Then again, dark matter might be axions. These were proposed by Roberto Peccei and Helen Quinn in 1977, as a way to solve the vexing CP problem. Some particle interactions violate a basic symmetry of nature, in which charge conjugation (C, converting a particle to its antiparticle) and parity (P, mirror reversal of space) are combined. This symmetry turns out not to be preserved in some particle interactions via the weak force. However, quantum chromodynamics, which involves the strong force, has CP symmetry. The question is, why? Peccei and Quinn resolved this difficulty by introducing an extra symmetry, broken by a new particle dubbed axion. Again, experimentalists have looked for them, but nothing convincing has been found.

If none of the above, what else?

Neutrinos are a wonderful example of bizarre particles that seemed almost impossible to detect. The Sun produces large numbers of them, but early detectors found only one third of the expected number of solar neutrinos. However, neutrinos come in three types, and it's now established that they transmute from one type to another as they travel. The early detectors could spot only one type. When upgraded to detect the others, the number tripled. Now, there could perhaps be a fourth type called a sterile neutrino. Standard-model neutrinos are left-handed; sterile ones, if they exist, are right-handed. (The jargon is chirality, which distinguishes particles from their mirror images.) If sterile neutrinos do exist they would bring neutrinos into line with all the other particles and also explain neutrino masses, which would be nice. They could possibly be dark radiation, which would mediate interactions between dark particles if they exist. Several experiments to detect them have been carried out. Fermilab's MiniBooNE found nothing in 2007, and the Planck satellite found nothing in 2013. But in a French experiment on neutrinos emitted from a nuclear reactor, 3% of the antineutrinos went AWOL. They might have been sterile neutrinos.

The catalogue of acronyms for experiments designed to detect dark matter, or likely to spot it, looks like a list of government-appointed quangos:[1] ArDM, CDMS, CRESST, DEAP, DMTPC, DRIFT, EDELWEISS, EURECA, LUX, MIMAC, PICASSO, SIMPLE, SNOLAB, WARP, XENON, ZEPLIN... Although these experiments have provided valuable data, and had many successes, they've not found any dark matter.

The Fermi Gamma-ray Space Telescope did spot a potential sign of dark matter at the heart of the Galaxy in 2010. Something was emitting a large quantity of gamma rays. This observation was seen as strong evidence for dark matter, some forms of which can decay into particles that produce gamma rays when they collide. Indeed, some physicists considered it to be a 'smoking gun' confirming the existence of dark matter. However, it now seems that the cause is ordinary matter: thousands of pulsars, which had gone unnoticed – not hard to miss, given the sheer quantity of objects in the cramped galactic core and the difficulties of observing that region. Moreover, if the gamma-ray excess really were caused by dark matter, other galaxies ought to emit similar quantities of gamma rays. According to Kevork Abazajian and

Ryan Keeley, they don't.[2] The smoking gun has turned out to be a damp squib.

In 2015 Gregory Ruchti, Justin Read, and others looked for different evidence of dark matter, in the disc of the Galaxy.[3] Over the aeons, the Milky Way has eaten dozens of smaller satellite galaxies, and should therefore have eaten their halos of dark matter too. As for a protoplanetary disc, this dark matter should be concentrated into a disc roughly coinciding with the Galaxy's ordinary matter. This can be detected, in theory, because it affects the chemistry of stars. The interlopers should be a bit hotter than the natives. However, a survey of 4675 candidate stars in the disc revealed nothing of the kind, although there were some such stars further out. It seems, therefore, that the Galaxy lacks a dark matter disc. That doesn't stop it having the conventional spherical halo, but it does add slightly to the worry that dark matter may not exist at all.

✦

Sometimes dark matter runs into trouble because there's too much of it. Recall that globular clusters are relatively small spheres of stars, which orbit our Galaxy and many others. Dark matter only interacts through gravity, so it can't emit electromagnetic radiation. Therefore it can't get rid of heat, a prerequisite for contraction under gravity, so it can't form clumps as small as globular clusters. Therefore globular clusters can't contain much dark matter. However, Scarpa finds that the stars in Omega Centauri, the largest of the Galaxy's globular clusters, move too fast to be explained by the visible matter. Since dark matter ought not to occur, something else, perhaps a different law of gravity, might be responsible for the anomaly.

Despite the expenditure of large amounts of ingenuity, time, energy, and money in a currently futile quest for dark matter particles, most astronomers, especially cosmologists, consider the existence of dark matter to be a done deal. Actually, dark matter doesn't perform as well as is usually claimed.[4] A spherical halo of dark matter, the standard assumption, doesn't give a terribly convincing galactic rotation curve. Other distributions of dark matter would work better – but then you'd have to explain why matter that interacts only via gravity should be distributed in such a manner. This kind of difficulty tends to be swept under the carpet, and questioning the existence of dark matter is treated as a form of heresy.

Admittedly, inferring the existence of unseen matter by observing anomalies in the orbits of stars or planets is a method with a long and mostly distinguished history. It led to the successful prediction of Neptune. It got lucky with Pluto, where the calculation was based on assumptions that turned out not to be valid, but nonetheless an object was found close to the predicted location. It has revealed several small moons of the giant planets. And it confirmed relativity when applied to an anomaly in the precession of the perihelion of Mercury. Moreover, many exoplanets have been discovered by inference from the way they make their parent star wobble.

On the other hand, on at least one occasion this method had a much less distinguished outcome: Vulcan. As we saw in Chapter 4, the prediction of this non-existent world, supposedly orbiting closer to the Sun than Venus, was an attempt to explain the precession of the perihelion of Mercury by attributing the anomaly to perturbation by an unobserved planet.

In terms of these precedents, the big question is whether dark matter is a Neptune or a Vulcan. The overwhelming astronomical orthodoxy is that it's a Neptune. But if so, it's a Neptune that currently lacks one key feature: Neptune itself. Against the orthodox view we must set a growing conviction, especially among some physicists and mathematicians, that it's a Vulcan.

✦

Since dark matter seems remarkably shy whenever anyone actually looks for it, perhaps we should contemplate the possibility that there isn't any. The gravitational effects that led cosmologists to postulate its existence seem undeniable, so possibly we must look elsewhere for an explanation. We might, for example, mimic Einstein and seek a new law of gravity. It worked for him.

In 1983 Mordehai Milgrom introduced Modified Newtonian Dynamics (MOND). In Newtonian mechanics, the acceleration of a body is exactly proportional to the force applied. Milgrom suggested that this relationship might fail when the acceleration is very small.[5] In the context of rotation curves, this assumption can also be reinterpreted as a slight change to Newton's law of gravity. The implications of this proposal have been worked out in some detail, and various objections have been disposed of. MOND was often criticised for not being relativistic, but in 2004 Jacob Bekenstein formulated a relativistic generalisation, TeVeS (tensor–vector–scalar gravity).[6] It's always unwise to criticise a new proposal for allegedly lacking some feature, when you haven't bothered to look for it.

Galactic rotation curves are not the only gravitational anomaly that astronomers have found. In particular, some galactic clusters seem to be bound together more strongly than the gravitational field of the visible matter can explain. The strongest example of such anomalies (according to dark matter proponents) occurs in the Bullet Cluster, where two galaxy clusters are colliding. The centre of mass of the two clusters is displaced from that inferred from the densest regions of normal matter, and the discrepancy is said not to be compatible with any current proposal to modify Newtonian gravity.[7] However, that's not the end of the story, because in 2010 a new study suggested that the observations are also inconsistent with dark matter, as formulated in the standard ΛCDM cosmological model. Meanwhile Milgrom has argued that MOND can explain the Bullet Cluster observations.[8] It has long been accepted that MOND doesn't fully explain the dynamics of galactic clusters, but it deals with about half of the discrepancy otherwise accounted for by dark matter. The other half, Milgrom believes, is merely unobserved ordinary matter.

That's more likely than dark matter enthusiasts tend to admit. In 2011 Isabelle Grenier became concerned that the cosmological sums don't work. Forget dark matter and dark energy: about half of the *ordinary* (jargon: baryonic) matter in the universe is missing. Now her group has found quite a lot of it, in the form of regions of hydrogen, so cold that it emits no radiation we can detect from Earth.[9] The evidence came from gamma rays, emitted by molecules of carbon monoxide, associated with cosmic dust clouds in the voids between the stars. Where there's carbon monoxide, there's normally hydrogen, but it's so cold that only the carbon monoxide can be detected. Calculations suggest that a huge amount of hydrogen is being overlooked.

Not only that: the discovery shows that our current views wildly underestimate the amount of normal matter. Not by enough to replace dark matter by ordinary matter, mind you – but enough to require a rethink of all of the dark matter calculations. Such as the Bullet Cluster.

On the whole, MOND can explain most of the anomalous observations related to gravity – the majority of which are open to several conflicting interpretations in any case. Despite that, it hasn't found favour among cosmologists, who argue that its formulation is rather arbitrary. I don't personally see why it's more arbitrary than positing vast amounts of a radically new kind of matter, but I suppose the point is that you can keep your treasured equations that way. If you change the equations, you need evidence to support your choice of new equations, and 'they fit observations' fails to clinch the need for precisely *that* modification. Again, I'm not convinced by this argument, because the same goes for new kinds of matter. Especially when no one has ever detected it except by inference from its supposed effects on visible matter.

There's a tendency to assume there are only two possibilities: MOND or dark matter. However, our theory of gravity is not a sacred text, and the number of ways to modify it is huge. If we don't explore that possibility, we could be backing the wrong horse. And it's a bit unfair to compare the tentative results of a few pioneers of such theories to the vast effort that conventional cosmology and physics has put into the quest for dark matter. Cohen and I call this the 'Rolls-Royce problem': no new design of car will ever get off the ground if you insist that the first prototype must be better than a Roller.

There are other ways to avoid invoking dark matter. José Ripalda has investigated time-reversal symmetry in general relativity, and its effect on negative energies.[10] The latter are usually thought to be ruled out because it would cause the quantum vacuum to decay through the creation of particle/antiparticle pairs. He points out that such calculations assume that the process occurs only in forward time. If the time-reversed process, in which particle pairs mutually annihilate, is also taken into consideration, the net effect on the vacuum is zero. Whether energy is positive or negative is not absolute; it depends on whether the observer is travelling into the future or the past. This introduces a notion of two types of matter: future-pointing and past-pointing. Their interaction requires two distinct metrics rather than the usual one, differing in their sign, so this approach can be viewed as a modification of general relativity.

According to this proposal, an initially homogeneous static universe would undergo an accelerating expansion as gravity amplifies quantum fluctuations. This removes the need for dark energy. As for dark matter, Ripalda writes: 'One of the more intriguing aspects of "dark matter" is that its apparent distribution is different from the distribution of matter. How could this be if "dark matter" interacts gravitationally and follows the same geodesics as all matter and energy?' Instead, Ripalda uses an analogy with electrostatics to suggest that the presumed spherical halo of dark matter surrounding a galaxy is actually a spherical void in an otherwise all-pervading distribution of past-pointing matter.

It might even be possible to get rid of all three add-ons, and possibly the Big Bang itself. The theory developed by Ali and Das, which eliminates the initial singularity of the Big Bang and even permits an infinitely old universe, is one way to achieve this. Another, according to Robert MacKay and Colin Rourke, is to replace the usual model of a universe that's smooth on large scales by one that's clumpy on small scales.[11] This model is consistent with the present-day geometry of the universe – indeed, more so than the standard pseudo-Riemannian manifold – and no Big Bang is required. The matter distribution of the universe could be static, while individual structures such as galaxies would come and go over a cycle lasting around 10^{16} years. Redshift could be geometric, caused by gravity, rather than cosmological, caused by the expansion of space.

Even if this theory is wrong, it demonstrates that changing a few assumptions about the geometry of spacetime makes it possible to preserve the standard form of Einstein's field equations, throw out the three *dei ex machina* of inflation, dark energy, and dark matter, and derive behaviour that is reasonably consistent with observations. Bearing in mind the Rolls-Royce problem, it makes a fair case for considering more imaginative models, instead of becoming committed to radical and otherwise unsupported physics.

✦

I've been keeping up my sleeve a potential way to retain the usual laws of nature while throwing out dark matter altogether. Not because there are exotic alternatives, but because the calculation that seems to prove dark matter exists may be erroneous.

I say 'may be' because I don't want to oversell the idea. But mathematicians are starting to question the assumptions that go into the Kepler equation, and their results, albeit incomplete, show that there's a case to answer. In 2015 Donald Saari[12] analysed the mathematical arguments used by cosmologists to justify the presence of dark matter, and found evidence that Newton's laws may have been misapplied in their theory of galactic structure and rotation curves.

If so, dark matter is probably a Vulcan.

Saari's concern is solely about the logical structure of the standard mathematical model that astronomers use to derive the Kepler equation. His calculations cast doubt on whether this model is appropriate. It's a radical proposal, but Saari is an expert on the mathematics of the n-body problem and gravitation in general, so it's worth running through his reasoning. I'll spare you the detailed sums – if you want to check them, consult his paper.

Everything rests on the Kepler equation. This follows directly and correctly from one key modelling assumption. A realistic model of a galaxy would involve hundreds of billions of stars. Their planets and other small bodies can probably be neglected, but a literal model is an n-body problem in which n is 100 billion or more. Maybe reducing that figure wouldn't change the results too much, but as we saw in Chapter 9, even when $n = 3$ (indeed $2\frac{1}{2}$) the n-body problem is intractable.

Astronomers therefore make a modelling assumption, which, together with an elegant mathematical theorem, simplifies the galaxy into a single body. Then they analyse the motion of a star around this body to deduce the theoretical rotation curve from the Kepler equation. The assumption is that when viewed on galactic scales, galaxies look more like a continuous fluid – a star soup – than a discrete system of n bodies. In this 'continuum' setting, a lovely theorem proved by Newton applies. (He used it to justify treating spherical planets as point masses.) Namely, subject to some reasonable symmetry assumptions, the total force exerted inside and on any particular spherical shell is zero, while the force exerted externally is the same as it would be if all the matter inside the shell were condensed at the central point.

Consider a star in a galaxy – call it a test star – and imagine a spherical shell, with the same centre as the galaxy, that passes through it. The mass inside the shell is what I previously called 'the mass inside that radius'. Independently of what the stars inside this shell are doing, we can apply Newton's theorem to concentrate their total mass at the centre of the galaxy, without affecting the overall force experienced by the test star. The stars outside the shell exert no force at all, because the test star lies on that shell. So the motion of the test star around the galaxy reduces to a *two*-body problem: one star going round a very heavy point mass. The Kepler equation follows directly from this.

The symmetry assumption required to apply Newton's theorem is that all stars follow circular orbits, and stars at the same distance from the centre move with the same speed. That is, the dynamics has rotational symmetry. It's then easy to derive exact solutions of the equations of motion for a star soup. You can choose either the formula for the distribution of mass, or the formula for the rotation curve, and use Kepler's equation to deduce the other one. There's one constraint: the mass must increase as the radius grows.

The star soup model is therefore self-consistent, agrees exactly with Newtonian gravity, and obeys Kepler's equation. The underlying assumption of circular symmetry also seems to agree with observations. So we obtain a time-honoured model, based on clever and valid mathematics, and it makes the problem soluble. No wonder astronomers like it.

Unfortunately, it's mathematically flawed. It's not yet clear how serious the flaw is, but it's not harmless, and it could be fatal.

Two aspects of the model are questionable. One is the assumption of circular orbits for all stars. But the most significant is the continuum approximation – star soup. The problem is that smoothing out all the stars inside the shell eliminates an important part of the dynamics. Namely, *interactions* between stars close to the shell and the star whose rotational speed we're trying to calculate.

In a continuum model, it makes no difference whether the matter inside the shell is rotating or stationary. All that counts is the total mass inside the shell. Moreover, the force that this mass exerts on the test star always points towards the centre of the galaxy. The Kepler equation depends on these facts.

However, in a real *n*-body system, stars are discrete objects. If a second star passes very close to the test star, discreteness implies that it dominates the local gravitational field, and attracts the test star towards it. So this nearby star 'tugs' the test along with it. This *speeds up* the rotation of the test star around the galactic centre. Of course it also slows the passing star down, but that's quickly replaced by another star, following along behind it. This intuitive argument suggests that the Kepler equation underestimates rotational speeds at large distances. If so, that helps to explain the anomaly.

Here's a very simplified analogy. Think of a tiny ball bearing resting on top of a spinning wheel (with both confined to a plane and no gravity to disturb the ball bearing). If the wheel is a perfectly smooth circle, it has no effect on the ball bearing and might as well be stationary. A discrete n-body model, however, replaces the wheel by a toothed cogwheel. Now each tooth of the cog hits the ball bearing, giving it a kick in the direction of rotation. Smaller teeth don't eliminate the kick because there are more of them. So the limiting kick for very small teeth is not the same as the kick for no teeth at all, which is zero.

This argument isn't just a vague piece of handwaving. Saari does the sums to prove that a smoothed-out star soup does *not* adequately model an n-body distribution for large n. In particular, it ignores tugging. However, the overall effect of the tugging might be small, because real n-body dynamics is more complicated than the scenario just analysed. To assess the importance of the tugging effect, we have to use an exact n-body model for all of the stars inside the shell, to find their combined effect on the test star.

The best way to do this is to construct an n-body state that retains all of the key properties assumed for the star soup, aside from continuity. If this particular state changes the Kepler equation, we can be pretty confident that the cause is replacing n discrete bodies by a continuous star soup. Those key properties are a symmetric mass distribution, with each star moving in a circle and its acceleration pointing towards the centre of the galaxy.

Although in general we can't write down explicit solutions of n-body problems, there's one class of solutions for which this can be done, known as central configurations. In these special states, concentric rings of stars resembling a spider's web all rotate with the same angular velocity, as if the configuration were rigid. This idea goes back to an 1859 paper by James Clerk Maxwell about the stability of Saturn's rings, mentioned in Chapter 6 as proof that the rings can't be solid. Saari uses a similar idea to suggest that the star soup can't model galaxy dynamics correctly.

Central configurations are artificial, in the sense that no one would expect a real galaxy to have such a regular form. On the other hand,

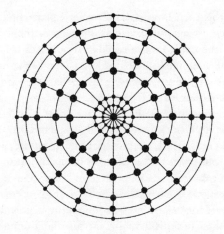

Central configuration. Any number of spokes and rings are possible.
The masses on each ring are equal but different rings can have
different masses. The radii of the rings can be adjusted. For a given
speed of rotation, masses can be chosen to suit given radii, or
conversely. Variations on the spiderweb theme are also possible.

they're a reasonable choice for testing how the continuum and *n*-body
models match up. If we choose enough radial lines in the web, and
enough circles, we get a very dense soup of stars, well approximated by
a continuum. The spiderweb configuration also satisfies, to a very good
approximation, the symmetry conditions used to derive the Kepler
equation. So the star soup approximation ought to work.

In particular, the Kepler equation should be valid for a spinning
spiderweb. We can check this using the version that expresses the
mass distribution in terms of the velocity at a given radius. Since the
spiderweb rotates rigidly, the velocity is proportional to the radius. The
Kepler equation therefore predicts a mass distribution proportional to
the cube of the radius. This result is valid whatever the masses of the
stars in the configuration actually are.

To check this, we now perform the *exact* discrete *n*-body calculation for the spiderweb. The theory of central configurations allows a
lot of flexibility in the choice of the masses of the stars. For example,
if each star (hence each ring) has the same mass, central configurations
exist, and the mass distribution is always less than a constant times

the radius. In this case, however, the Kepler equation tells us that the outermost ring has *one million times* the mass of the innermost one, when actually their masses are equal. So the exact calculation does *not* validate the simplified model leading to the Kepler equation. On the contrary, as the radius increases, the correct mass grows far more slowly than the Kepler formula predicts.

This calculation proves that the star soup model can produce seriously incorrect results, even when the assumptions on which the model is based are satisfied. Despite the popular phrase, which misuses the obsolete term 'prove', it takes only one exception to *disprove* the rule.[13]

Saari's calculations have another important consequence. If dark matter exists, and forms vast, massive halos around galaxies, as astronomers think, then it can't actually explain the anomalous rotation curve that started the whole business. Either the law of gravity, or the traditional modelling assumptions, must be wrong.

Outside the Universe

Sometimes the Star Maker flung off creations which were in effect
groups of many linked universes, wholly distinct physical systems
of very different kinds.

Olaf Stapledon, *Star Maker*

WHY ARE WE HERE?

It's the ultimate philosophical question. Humans look through the
windows of their eyes at a world far bigger and more powerful than
they. Even when the only world you know is a small village in a forest
clearing, there are thunderstorms, lions, and the occasional hippopot-
amus to contend with, and those are already awe-inspiring. When, as
cosmologists now think, your world is 91 billion light years across and
growing, it's downright humbling. There's an awful lot of 'here' and
very little 'we'. Which makes it a very big 'why'.

However, humanity's sense of self-importance never stays humbled
for long. Fortunately, neither does its sense of wonder or its insatiable
curiosity. And so we dare to ask the ultimate question.

The objections discussed in the previous two chapters have not
dented cosmologists' conviction that they know the answer: the Big
Bang and its add-ons correctly describe how the universe came into
being. Physicists are similarly convinced that relativity and quantum
theory together explain how the universe behaves. It would be good to
unify these theories, but they generally work well on their own if you
choose the right one.

Biology tells an even more convincing story of the origin of life
and its evolution into the millions of species that inhabit the Earth

today, us among them. Some devotees of some belief systems claim that evolution requires incredibly unlikely coincidences, but biologists have repeatedly explained the flaws in those arguments. Our understanding of life on Earth has many gaps, but one by one they're being filled in. The main story hangs together, supported by at least four independent lines of evidence – the fossil record, DNA, cladistics (family trees of organisms), and breeding experiments.

However, when it comes to cosmology, even cosmologists and physicists are worried that the universe as we understand it requires some pretty enormous coincidences. The issue is not explaining what the universe does; it's why that particular explanation is valid, rather than a host of others that on the face of it seem equally likely. This is the problem of cosmological fine-tuning, and creationists and cosmologists alike take it very seriously.

Fine-tuning comes into play because physics depends on a number of fundamental constants, such as the speed of light, Planck's constant in quantum theory, and the fine structure constant, which determines the strength of the electromagnetic force.[1] Each constant has a specific numerical value, which scientists have measured. The fine structure constant is about 0·00729735, for example. No accepted physical theory predicts the values of these constants. For all we know, the fine structure constant could have been 2·67743, or 842,006,444·998, or 42.

Does it matter? Very much. Different values for the constants lead to different physics. If the fine structure constant were a bit smaller or a bit bigger, atoms would have a different structure and might even become unstable. So there wouldn't be people, or a planet for them to live on, or atoms to make them from.

According to many cosmologists and physicists, the values of the constants that make people *possible* have to be within a few per cent of the values in this universe. The chance of that happening for just *one* constant is about the same as tossing six heads in a row. Since there are at least 26 constants, the chance that our universe has the values that it does, and is thus suitable for life to exist, is like tossing 156 heads in a row. Which is about 10^{-47}, or

0·000 000 000 000 000 000 000 000 000 000 000 000 000 000 000 01

So, basically, we shouldn't be here at all.

Yet … here we are. And that's a puzzle.

Some religious people see this calculation as a proof of the existence of God, who has the luxury of selecting values of the fundamental constants that make life possible. However, a God with that sort of power could equally well have chosen totally different fundamental constants and then worked a miracle, so the universe would exist anyway despite having the wrong constants. Indeed, there's no reason for an omnipotent creator to use fundamental constants at all.

There seem to be two options. Either some supernatural power arranged it, or future physics will explain the apparent coincidences and show why the existing fundamental constants are inevitable.

Recently, cosmologists added a third: the universe tries all possible values in turn. If so, the universe will eventually stumble upon numbers suitable for life, and life will evolve. If intelligent life appears, and its understanding of cosmology grows, it will get really puzzled about why it's there. Only when it thinks of the third option will it stop worrying.

The third option is called the multiverse. It's fresh, it's original, and you can do some really clever physics with it. I'll spend most of this chapter on various versions of it.

Then I'll give you option four.

✦

Modern cosmology has put together what it believes is a pretty accurate description of what we normally mean by 'universe', so a new word has been invented: multiverse. This comprises the universe in its usual sense, plus any number of hypothetical extras. These 'parallel' or 'alternative' (US: alternate) worlds may coexist with ours, be located outside it, or be entirely independent of it. These speculations are often dismissed as unscientific, because it's difficult to test them against real data. However, some are testable, at least in principle, by the standard scientific method of inferring things that can't be seen or measured directly from things that can. Comte thought it impossible ever to know the chemical composition of a star. Spectroscopy turned this belief on its head: often the chemical composition of a star is almost all we know about it.

In *The Hidden Reality*,[2] mathematical physicist Brian Greene describes nine different types of multiverse. I'll discuss four:

- *Quilted multiverse*: an infinite patchwork in which any region has a nearly exact copy somewhere else.

- *Inflationary multiverse*: whenever eternal inflation blows up the television and the cat, a new universe with different fundamental constants bubbles off.

- *Landscape multiverse*: a network of alternative universes linked by quantum tunnelling, each obeying its own version of string theory.

- *Quantum multiverse*: a superposition of parallel worlds, each with its own separate existence. This is a universe version of Schrödinger's celebrated cat, both alive and dead at the same time.

Greene argues that it's sensible to consider these alternative universes, and explains how they're to some extent supported by modern physics. Additionally, various issues that we don't understand can be resolved by multiverse thinking. He points out that fundamental physics has repeatedly shown that the naive view of the world presented by our senses is wrong, and we can expect this state of affairs to continue. And he places some weight on the common feature of multiverse theories: they all 'suggest that our common-sense picture of reality is only part of a grander whole'.

I'm not convinced that having lots of mutually inconsistent speculations makes any one of them more likely to be true. It's like religious sects: unless you're a true believer, fundamental differences in doctrine allied to common claims of divine revelation tend to discredit them all. But let's take a look at a few multiverses, and you can decide for yourself. Naturally, I'll throw in some thoughts of my own.

✦

I'll start with the quilted multiverse. Actually, it's not really a multiverse, but a universe so big that its inhabitants can observe only patches. But the patches overlap. It depends on space being infinite, or at least unimaginably vast – way bigger than the observable universe. When this idea is combined with the discrete nature of quantum mechanics, it has an interesting consequence. The number of possible quantum states for the observable universe, though gigantic, is finite. The observable universe can do only finitely many different things.

For simplicity, consider an infinite universe. Conceptually cut it up into pieces, like a patchwork quilt, so that each piece is big enough to contain the observable universe. Patches of equal size have the same number of possible quantum states: I'll call these patch-states. Since an infinite universe contains infinitely many patches, each with the same finite number of states, at least one patch-state must occur infinitely often.[3] Taking into account the random nature of quantum mechanics, *every* patch-state is certain to occur infinitely often.

The number of distinct patch-states for patches the size of the observable universe is about $10^{10^{122}}$. That is, write down 1 followed by 122 zeros; then start again, writing down 1 followed by *that* gigantic number of zeros. (Don't try this at home. The universe contains too few particles to make the ink or the paper, and it would end soon after you got started.) By similar reasoning, the nearest exact copy of *you* is about $10^{10^{128}}$ light years away. For comparison, the edge of the observable universe is less than 10^{11} light years away.[4]

Inexact copies are easier to arrange, and more interesting. There might be a patch that contains a copy of you, except your hair colour is different, or you're a different sex, or you live next door, or in another country. Or you're the prime minister of Mars. These near misses are more common than exact copies, although they're still exceedingly few and far between.

We can't visit regions a few light years away, let alone $10^{10^{128}}$ light years, so it seems impossible to test this theory scientifically. The definition of a patch rules out causal connections between patches that don't overlap, so you can't get there from here. Possibly some theoretical consequence could be tested, but it's a tenuous hope, and it would depend on the theory upon which the inference was based.

✦

The landscape multiverse is especially interesting because it could solve the vexing cosmological conundrum of fine-tuning.

The idea is simple. The chance of *any particular* universe having just the right fundamental constants may be small, but that's no obstacle if you make enough universes. With odds of 10^{47} to one against, you stand a reasonable chance of getting one universe suited for life by making 10^{47} of them. Make even more, and the likely number of successes gets bigger. In any such universe – and only those – life can originate, evolve, get round to asking 'why are we here?', discover how improbable that is, and start worrying about it.

At first sight this is like the weak anthropic principle: the only universes in which creatures can ask 'why are we here?' are those that make being here possible. The general view is that this fact alone doesn't fully resolve the difficulty. It raises the question: if there's only one universe, how come it made such an improbable choice? However, in the context of the landscape multiverse, that's not an issue. If you make enough random universes, life in at least one of them becomes a dead cert. It's a bit like the lottery. The chance that Mrs Smith down the road will win the lottery on any given draw is (in the UK, until recent changes) about one in 14 million. However, millions of people play the lottery, so the chance that *someone* will win is much larger, about two chances out of three. (One third of the time no one wins, and there's a 'rollover' where the prize is added to the pot for the next draw.)

In the landscape multiverse, life wins the cosmological lottery by buying every ticket.

Technically, the landscape multiverse is a string-theoretic variant of the inflationary multiverse. String theory is an attempt to unify relativity with quantum mechanics by replacing point particles by tiny multidimensional 'strings'. This isn't the place to go into details, but there's a big problem: there are about 10^{500} different ways to set up string theory.[5] Some produce fundamental constants much like those in our universe; most don't. If there were some magical way to single out a particular version of string theory, we could predict the fundamental constants, but right now there's no reason to prefer one version to any other.

A string-theoretic multiverse allows them all to be explored, one at a time – a bit like serial monogamy. If you wave your theorist's hands hard enough, quantum uncertainty might allow occasional transitions from one version of string theory to another, so *the* universe performs a drunkard's walk through the space of all string-theoretic universes. The constants are close to ours, so life can evolve. As it happens, those fundamental constants also produce very long-lived universes with features like black holes. So the serially changing universe tends to hang around the interesting localities – where beings like us hang out.

That raises a subtler question. How come suitability for life and longevity are associated? Lee Smolin has suggested an answer for the inflationary multiverse: new universes that bud off via black holes might evolve by natural selection, homing in on combinations of fundamental constants that not only make life possible, but give it plenty of time to get started and become more complex. It's a nice idea, but it's not clear how two universes can compete with each other to bring Darwinian selection into play.

The landscape multiverse has a certain amount going for it, but in the words of Lewis Carroll, it's 'a maxim tremendous but trite'.[6] It can explain *anything*. A seven-tentacled metalloid cyberorganism living in a universe with totally different fundamental constants could advance exactly the same reason why *its* universe exists, and is so finely tuned for metalloid cyberlife. When a theory predicts every possible outcome, how can you test it? Can it really be considered scientific?

George Ellis has long been a multiverse sceptic. Writing about the inflationary multiverse, but adding that similar remarks apply to all types, he said:[7]

> The case for the multiverse is inconclusive. The basic reason is the extreme flexibility of the proposal ... so we are supposing the existence of a huge number – perhaps even an infinity – of unobservable entities to explain just one existing universe. It hardly fits 14th-century English philosopher William of Ockham's stricture that 'entities must not be multiplied beyond necessity'.

He ended on a more positive note: 'Nothing is wrong with scientifically based philosophical speculation, which is what multiverse proposals are. But we should name it for what it is.'

✦

The quantum multiverse is the oldest around, and it's all Erwin Schrödinger's fault. The cat, right? You know, the one that's both alive and dead at the same time until you look to see which. Unlike the other multiverses, the different worlds of the quantum multiverse coexist with each other, occupying the same space and time. Science fiction writers love it.

Independent coexistence is possible because quantum states can be *superposed*: added together. In classical physics, water waves do something similar: if two wave trains cross each other, their peaks add together to create bigger peaks, but a peak and a trough cancel out. However, this effect goes much further in the quantum realm. For example, a particle may spin clockwise or anticlockwise (I'm simplifying here but you get the idea). When these states are superposed, they *don't* cancel out. Instead, you get a particle that is spinning both ways at once.

If you make a measurement when the system is in one of these combined states, something remarkable happens. You get a definite answer. This led to a lot of argument among the early pioneers of quantum theory, resolved at a conference in Denmark when most of them agreed that the act of observing the system somehow 'collapses' the state to one or other component. It's called the Copenhagen interpretation.

Schrödinger wasn't entirely convinced, and he invented a thought experiment to explain why. Place a cat in an impermeable box, along with a radioactive atom, a bottle of poison gas, and a hammer. Rig up some machinery so that if the atom decays, emitting a particle, the hammer smashes the bottle and the gas kills the cat. Shut the box, and wait.

After some time, you ask: is the cat alive or dead?

In classical (that is, non-quantum) physics, it's either one or the other, but you can't decide which until you open the box. In quantum physics, the state of the radioactive atom is in a superposition of 'decayed' and 'not decayed', and it remains like that until you observe the state by opening the box. Then the atom's state immediately collapses to one or the other. Schrödinger pointed out that the same therefore applies to the cat, which can be thought of as a huge system of interacting quantum particles. The machinery inside the box ensures that the cat remains alive if the atom hasn't decayed, but dies if it has. So the cat must be alive and dead at the same time ... until you open the box, collapse the cat's wavefunction, and find out which.

In 1957 Hugh Everett applied similar reasoning to the universe as a whole, suggesting that this could explain how the wavefunction collapses. Bryce DeWitt later called Everett's proposal the many-worlds interpretation of quantum mechanics. Extrapolating from the cat, the universe itself is a combination of all of its possible quantum states. However, this time there's no way to open the box, because there's nothing outside the universe. So nothing can collapse the universe's quantum state. But an internal observer is part of one of its component states, and therefore sees only the corresponding part of the wavefunction of the universe. The live cat sees an atom that hasn't decayed, while the parallel dead cat... Hmm, let me give that some more thought.

In short, each parallel observer sees itself as inhabiting just one of a huge number of parallel worlds, all coexisting but in different states. Everett visited Niels Bohr in Copenhagen to tell him about this idea, but Bohr was outraged by the proposal that the quantum wavefunction of the universe does not and cannot collapse. He and his like-minded colleagues decided that Everett had no understanding of quantum mechanics, and said so in uncomplimentary terms. Everett described the visit as 'doomed from the beginning'.

It's a very curious idea, even though it can be formulated in a mathematically sensible way. It doesn't help that the many-worlds interpretation is often expressed in terms of historical events in a misguided attempt to make it comprehensible. In the component universe that you and I are observing, Hitler lost World War II. But there's another parallel universe in which he (well, actually another Hitler altogether, though no one says so) won the war (well, a different war ...), and those versions of you and me perceive themselves as living in that one. Or maybe we died in the war, or never got born ... who knows?

Plenty of physicists insist that the universe *really is like that*, and they can prove it. Then they tell you about experiments on electrons. Or, more recently, molecules. But Schrödinger's intention was to point out that a cat is not an electron. Viewed as a quantum-mechanical system, a cat consists of a truly gigantic number of quantum particles. Experiments on a single particle, or a dozen, or even a billion, don't tell us anything about a cat. Or a universe.

Schrödinger's cat went viral among physicists and philosophers, generating a huge literature with all sorts of supplementary questions. Why not put a movie camera inside too, film what happens, then look at it afterwards? But no, that won't work: until you open the box, the camera will be in a combination of 'filmed dead cat' and 'filmed live cat'. Can't the cat observe its own state? Yes if it's alive, no if it's dead – but an external observer still has to wait for the box to be opened. Give the animal a mobile phone – no, this is getting silly, and *that* would superpose too. Anyway, it's an impermeable box. It has to be, or you could infer the state of the cat from the outside.

Impermeable boxes don't actually exist. How valid is a thought experiment about an impossibility? Suppose we replace that radioactive atom by an atomic bomb, which either explodes or not. By the same argument, until we open the box, we don't know which it did. The military would kill for a box that remains undisturbed when you set off a nuclear weapon inside it.

Some go further, claiming that only a human (or at least an intelligent) observer will do – a massive slur on the feline race. Some suggest that the reason the universe brought us into existence is so that we can observe it, thus collapsing its wavefunction and bringing *it* into existence. We're here because we're here because we're here.

✦

This remarkable causal inversion elevates the importance of humanity, but it ignores the feature that led Bohr to dismiss Everett's theory: in the many-worlds interpretation the wavefunction of the universe *doesn't* collapse. It flies in the face of the Copernican principle and smacks of hubris. It also misses the point: the puzzle of Schrödinger's cat is about observations, not observers. And it's not really about what happens when an observation is made. It's about what an observation *is*.

The mathematical formalism of quantum mechanics has two aspects. One is Schrödinger's equation, which is used to model quantum states, and has well-defined mathematical properties. The other is how we represent an observation. In the theory, this is a mathematical function. You put a quantum system into the function, and its state – the result of the observation – pops out of the other end. Just as you insert the number 2 into the logarithm function and log 2 emerges. This is all very neat and tidy, but what actually happens is that the state of the system interacts with the state of the measuring apparatus, which is a far more complex quantum system. That interaction is too complicated to study mathematically in any detail, so it's assumed that it reduces to a single, tidy function. But there's no reason to suppose that's actually the case, and every reason to suspect that it's not.

What we have is a mismatch between an exact but intractable quantum representation of the measuring process, and an *ad hoc* add-on, the hypothetical function. No wonder strange and conflicting interpretations emerge. Similar issues turn up all over quantum theory, largely unnoticed. Everyone is focusing on the equations and how to solve them; nobody thinks about the 'boundary conditions' that represent apparatus or observations.

A nuclear-bomb-proof box is a case in point. Another example is the half-silvered mirror, which reflects some light while letting the rest pass straight through. Quantum experimentalists like this gadget because it acts as a beam-splitter, taking a stream of photons and diverting them randomly in two different directions. After they've done whatever you want to test, you recombine them again to compare what happened. In quantum-mechanical equations, a half-silvered mirror is a crisp object that has no effect on a photon except to redirect it at right angles with 50% probability. It's like a cushion on a pool table that sometimes makes the ball bounce, perfectly elastically, and sometimes disappears and lets it pass straight through.

However, a real half-silvered mirror is a huge quantum system composed of silver atoms scattered on a sheet of glass. When a photon hits the mirror, it either bounces off a subatomic particle in a silver atom, or it penetrates further. It can bounce in any direction, not just at right angles. The layer of silver atoms is thin, but thicker than a single atom, so the photon may hit a silver atom deeper down, never mind the very messy atomic structure of the glass. Miraculously, when all these interactions are combined the photon is either reflected or passes through unchanged. (Other possibilities exist, but they're so rare we can ignore them.) So the reality is not like the pool ball. It's more like driving a photon car into a city from the north, and letting it interact with thousands of other cars, after which it miraculously emerges either to the south or to the east, at random. This complicated system of interactions is ignored in the clean, tidy model. All we have then is a fuzzy photon and a crisp, though randomly reflecting, mirror.

Yes, I know it's a model, and it seems to work. But you can't keep throwing this kind of idealisation into the pot while maintaining that all you're using is Schrödinger's equation.

✦

More recently, physicists have been thinking about quantum

observations from a genuinely quantum-mechanical point of view, instead of postulating unrealistic classical-type constraints. What they've found puts the whole matter in a far more sensible light.

First, I must acknowledge that cat-like superpositions of states have been created in the laboratory for increasingly large quantum systems. Examples, in approximate order of size, include a photon, a beryllium ion, a molecule of buckminsterfullerene (60 carbon atoms arranged to form a truncated icosahedral cage), and an electrical current (composed of billions of electrons) in a SQUID (superconducting quantum interference device). A piezoelectric tuning fork, made from trillions of atoms, has been placed in a superposition of vibrating and non-vibrating states. Not yet cats, but remarkable and counterintuitive. Getting closer to living creatures, Oriol Romero-Isart and colleagues proposed in 2009 to create a Schrödinger's flu virus.[8] Place a virus in a vacuum, cool it to its lowest-energy quantum state, and zap it with a laser. The influenza virus is hardy enough to survive such treatment, and it should end up in a superposition of that state and an excited higher-energy state.

This experiment hasn't been performed yet, but even if someone made it work, a virus isn't a cat. Quantum states of large-scale objects differ from those of small-scale objects such as electrons and SQUIDS, because superpositions of states of large systems are far more fragile. You can put an electron in a combination of clockwise and anticlockwise spin and keep it there almost indefinitely by isolating it from the outside world. If you try that with a cat, the superposition decoheres: its delicate mathematical structure quickly falls apart. The more complex the system, the faster it decoheres. The upshot is that even in a quantum model, a cat behaves like a classical object unless you look at it for an unobservably short time. The fate of Schrödinger's cat is no more mysterious than not knowing what's in your Christmas present from Auntie Vera until you unwrap it. Yes, she always sends either socks or a scarf, but that doesn't mean her present is a superposition of the two.

Dissecting the universe's quantum wavefunction into superpositions of human narratives – Hitler did or did not win – was always nonsense. Quantum states don't tell human stories. If you could look at the quantum wavefunction of the universe, you wouldn't be able to pick out Hitler. Even the particles that make him up would keep changing as he shed hair or picked up dust on his greatcoat. Similarly, there's no way to tell from the quantum wavefunction of a cat whether it's alive, dead, or has just turned into a cactus.

✦

Even within the framework of quantum mechanics, there's also a mathematical issue with the usual approach to the Schrödinger's cat paradox. In 2014 Jaykov Foukzon, Alexander Potapov, and Stanislaw Podosenov[9] developed a new, complementary approach. Their calculations indicate that even when the cat *is* in a superposed state, the state observed when the box is opened has 'definite and predictable measurement outcomes'. They conclude that: 'Contrary to [other] opinions, "looking" at the outcome changes nothing, beyond informing the observer of what has already happened.' In other words, the cat is definitely either alive or dead before anyone opens the box – but an external observer doesn't know which at that stage.

The core of their calculation is a subtle distinction. The usual representation of the superposed state of the cat is

$$|cat\rangle = |alive\rangle + |dead\rangle$$

Here the symbols $|\ \rangle$ are how quantum physicists write a particular kind of state,[10] so you can read it as 'state of'. I've left out some constants (probability amplitudes) by which the states are multiplied.

However, this formulation is inconsistent with the time evolution of quantum states. The Ghirardi–Rimini–Weber model, a mathematical technique for analysing the collapse of the wavefunction,[11] requires the explicit introduction of time. Causality forbids combining states that occur at different times, so we must write the state as

$$|cat\ at\ time\ t\rangle = |live\ cat\ at\ time\ t\ and\ undecayed\ atom\ at\ time\ t\rangle$$
$$+ |dead\ cat\ at\ time\ t\ and\ decayed\ atom\ at\ time\ t\rangle$$

This is an *entangled* state, in quantum jargon. It's not a superposition of 'pure' states such as 'live cat' or 'undecayed atom'. Instead, it's a superposition of mixed states, cat-state *and* atom-state, which represents the collapsed state of the coupled cat/atom *system*. It tells us that before we opened the box, either the atom had already decayed *and* (entirely predictably) killed the cat, or it hadn't and didn't. Which is what you'd expect from a classical model of the observational process, and isn't paradoxical.

In 2015 Igor Pikovski, Magdalena Zych, Fabio Costa and Časlav Brukner introduced a new ingredient, finding that gravity makes superpositions decohere even faster. The reason is relativistic time dilation – the effect that causes time to freeze at the event horizon of a black hole. Even the very tiny time dilation caused by a weak gravitational field interferes with quantum superpositions. So gravity decoheres Schrödinger's cat almost instantly into either 'alive' or 'dead'. Unless you assume the box is impervious to gravity, which is tricky since no such material exists.

There are probably more points of view on Schrödinger's cat, and the closely associated many-worlds interpretation of quantum mechanics, than there are quantum physicists. I've discussed just a few attempts to resolve the paradox, which suggest that the quantum multiverse is by no means a done deal. So you can stop worrying that somewhere parallel to this universe is another one in which another you is living in a world where Hitler triumphed. It might be *possible*, but quantum mechanics provides no compelling reasons to believe it's true.

But it *is* true for a photon. And even that is remarkable.

✦

By now you'll have realised that I'm a bit sceptical about multiverses. I love their mathematics, and they set up imaginative science fiction plots, but they involve too many unsupported assumptions. Among the various versions I've discussed, the landscape multiverse stands out as having something going for it. Not because we have evidence that it really exists – whatever that means – but because it seems to resolve the vexed issue of incredibly unlikely fine-tuning of fundamental constants.

Which brings me at last to option four.

The landscape multiverse is philosophical overkill. It tries to resolve a single question that currently puzzles a few cosmically insignificant humans by postulating an extraordinarily vast and complex object that transcends human experience entirely. It's like geocentric cosmology, in which the rest of the vast universe spins once a day round a fixed, central Earth. The physicist Paul Steinhardt, who worked on inflation at its inception, made a similar point about the inflationary multiverse:[12] 'In order to explain the one simple universe we can see, the inflationary multiverse hypothesis posits an infinite variety of universes with arbitrary amounts of complexity that we cannot see.'

It would be simpler to admit that we don't know why fine-tuning happens. But maybe we don't even need to go that far, because there's another possibility. Namely, the problem of fine-tuning has been massively exaggerated, and it doesn't actually exist. That's option four. If it's correct, multiverses are superfluous froth.

The reasoning is based on a more careful analysis of the alleged evidence for fine-tuning, that 10^{-47} chance of getting a combination of fundamental constants that's suitable for life. The calculation requires some strong assumptions. One is that the only way to make a universe is to choose 26 constants to insert into our current equations. It's true that mathematically these constants act as numerical 'parameters', modifying the equations without affecting their general mathematical form, and as far as we can tell, each modification gives a viable set of equations that define a universe. But we don't actually know that. We've never observed a modified universe.

Being a mathematician, I worry about a whole lot of other parameters that are tacitly included in the equations but never get written down because in our universe they happen to be zero. Why can't those vary too? In other words, what about putting extra terms into the equations, different from the ones we currently write down? Every extra term of this kind introduces yet more fine-tuning to explain. Why does the state of the universe *not* depend on the total number of sausages sold in Smithfield market in London in 1997? Or on the third derivative of the karmabhumi field, as yet unknown to science?

Gosh, two more constants whose values have to be very, very close to what happens in this universe.

It's remarkably unimaginative to think that the only way to make alternative universes is to vary the known fundamental constants in the currently fashionable model equations. It's like the inhabitants of a south sea island in the sixteenth century imagining that the only way to improve agriculture is to cultivate a better type of coconut.

However, let's give fine-tuning devotees the benefit of the doubt, and take this particular assumption for granted. Surely *then* that 10^{-47} comes into play, and demands explanation? To answer that question, we need to know a bit more about the calculation. Broadly speaking, the method is to fix all of the fundamental constants except one, and find out what happens when that particular constant changes. You then take some important real-world phenomenon, such as an atom, and see what effect the new value for that constant would have on the standard description of the atom. Lo and behold, the usual mathematics of atoms falls apart unless the change in the constant is very small.

Now do the same for another fundamental constant. Maybe this one affects stars. Leave all the other constants at their value in this universe, but change that one instead. This time you find that conventional models of stars cease to work unless the change in *that* constant is very small. Putting it all together, changing any constant by more than a very tiny amount causes *something* to go wrong. Conclusion: the only way to get a universe with the important features of this one is to use pretty much the same constants as this one. Do the sums, and out pops 10^{-47}.

It sounds convincing, especially if you look at all the impressive physics and mathematics involved in the sums. In *The Collapse of Chaos* in 1994, Cohen and I presented an analogous argument where the conceptual error is more obvious. Think of a car – say a Ford Fiesta. Think of some component, say the bolts that hold the engine together, and ask what happens if you change the diameter of the bolts *while leaving everything else fixed*. Well, if the bolts are too thick, they won't go in; if they're too thin, they'll be loose and fall out. Conclusion: to obtain a viable car, the diameter of the bolts has to be very close to what you find in a Ford Fiesta. The same goes for the wheels (change their size and the tyres don't fit), the tyres (change their size and the wheels don't fit), the spark plugs, each individual cogwheel in the gears, and so on. Put all this together and the chance of selecting parts that can make a car is much smaller than 10^{-47}. You can't even make a wheel.

In particular, there's only one possible car, and it has to be a Ford Fiesta.

Now stand at the corner of the street and watch all those Volkswagens, Toyotas, Audis, Nissans, Peugeots, and Volvos going past.

Something is clearly wrong.

✦

The mistake is to consider varying the constants *one at a time*.

If you're building a car, you don't start with a design that works, and then change all the bolt sizes while leaving the nuts the same as before. Or change the tyre sizes while leaving the wheels fixed. That's crazy. When you change the specification of one component, there are automatic knock-on effects on other components. To get a new design for a car that works, you make coordinated changes of *many* numbers.

I've encountered one response to this criticism of fine-tuning that boils down to 'Oh, but doing the sums is much harder if you vary several constants.' Yes, it is. But that doesn't justify doing a simpler sum if it's the *wrong* sum. If you went into a bank and asked for the balance in your account, and the clerk said 'Sorry, it's much too hard to find *your* balance, but Mrs Jones has £142', would you be satisfied?

The calculations used for fine-tuning also tend to ignore a very interesting and important question: if conventional physics doesn't work when you change some constants, what happens *instead*? Perhaps something else can play a similar role. In 2008 Fred Adams examined this possibility for a central part of the problem, star formation.[13] (Of course, stars are only part of the process that equips a universe with intelligent life forms. Victor Stenger tackles several others in his carefully argued *The Fallacy of Fine-Tuning*.[14] The outcome is the same: fine-tuning is a massive exaggeration.) Only three constants are significant for star formation: the gravitational constant, the fine structure constant, and a constant that governs nuclear reaction rates. The other 23, far from requiring fine-tuning, can take any value without causing trouble in that context.

Adams then looked at all possible combinations of those three important constants, to find out when they produce workable 'stars'. There's no reason to limit the definition of a star to the exact features that occur in our universe. You wouldn't be impressed by fine-tuning if someone told you it predicted stars can't exist, but that slightly hotter objects 1% larger, looking remarkably similar to stars, can. So Adams defines a star to be any object that holds together under its own gravity, is stable, survives for a long time, and uses nuclear reactions to produce energy. His calculations demonstrate that stars in that sense exist for a huge range of constants. If the universe maker chooses constants at random, there's a 25% probability of getting a universe that can make stars.[15]

That's not fine-tuning. But Adams's results are even stronger. Why not allow more exotic objects to count as 'stars' too? Their energy output could still sustain a form of life. Maybe the energy comes from quantum processes in black holes, or clumps of dark matter that generate energy by annihilating ordinary matter. Now the probability increases to 50%. As far as stars go, our universe isn't battling odds of 10 million trillion trillion trillion to one against. It just called 'heads', and the fundamental-constant coin landed that way up.

Epilogue

The universe is a big place, perhaps the biggest.
            Kilgore Trout (Philip José Farmer), *Venus on the Half Shell*

OUR MATHEMATICAL JOURNEY has taken us from the surface of the Earth to the outermost reaches of the cosmos, and from the beginning of time to the end of the universe. It began in the depths of prehistory, when early humans looked at the night sky and wondered what was going on up there. Its end is not yet in sight, for the more we learn about the cosmos, the more we fail to understand.

Mathematics has evolved alongside astronomy and related areas such as nuclear physics, astrophysics, quantum theory, relativity, and string theory. Science asks questions, mathematics attempts to answer them. Sometimes it's the other way round, and mathematical discoveries predict new phenomena. Newton's effort to formulate the laws of gravity and motion motivated the theory of differential equations and the n-body problem; these in turn inspired calculations predicting the existence of Neptune and the chaotic tumbling of Hyperion.

As a result, mathematics and science – astronomy in particular – have become more sophisticated as each inspires new ideas in the other. Babylonian records of planetary motion required high-precision arithmetic. Ptolemy's model of the solar system was based on the geometry of spheres and circles. Kepler's version required Greek geometers' conic sections. When Newton reformulated the whole business as a universal law, he presented it using complicated geometry, but his thinking was informed by calculus and differential equations.

The differential equation approach turned out to be better suited to the complexities of astronomical phenomena. With the motion of two mutually gravitating bodies understood, astronomers and mathematicians tried to move on to three or more. This attempt was stymied by what we now see as chaotic dynamics; indeed, it was in the $2\frac{1}{2}$-body problem that chaos first raised its head. But progress was still possible. Poincaré's ideas inspired an entirely new area of mathematics: topology. He himself was prominent in its early development. Topology is geometry in a very flexible sense.

The simple question 'how does the Sun shine?' opened up Pandora's box when it was realised that if it uses a conventional source of energy it should have burned to a cinder long ago. The discovery of nuclear physics explains how stars produce heat and light, culminating in accurate predictions of the abundances in the Galaxy of almost all chemical elements.

The dynamics of galaxies, with their striking shapes, inspired new models and insights, but also threw up a gigantic puzzle: rotation curves that disagree with Newton's law of gravity unless (as cosmologists claim) most of the matter in the universe is completely different from anything we've ever observed or created in particle accelerators. Or maybe, as some mathematicians are starting to wonder, the problem lies not with the physics, but with an inappropriate mathematical model.

When Einstein created a revolution in physics, and wanted to extend it to gravity, another type of geometry came to his rescue: Riemann's theory of manifolds, arising from Gauss's radical approach to curvature. The resulting theory of general relativity explains the anomalous precession of the perihelion of Mercury and the bending of light by the Sun. When it was applied to massive stars, strange mathematical features of the solutions drew attention to what we now refer to as black holes. The universe was starting to look very strange indeed.

When general relativity was applied to the universe as a whole, it looked even stranger. Hubble's discovery of redshifted galaxies, implying that the universe is expanding, led Lemaître to his exploding cosmic egg, aka the Big Bang. Understanding the Big Bang required new physics and mathematics, and powerful new computational methods. What at first sight looked like a complete answer started to fall apart as further data came in, requiring three different add-ons: inflation, dark matter, and dark energy. Cosmologists promote these as profound discoveries, which is true if their theories survive scrutiny; however, each add-on has its own problems, and none is supported by independent confirmation of the far-reaching assumptions required to make it work.

Scientists are constantly refining their understanding of the cosmos, and each new discovery raises fresh questions. In June 2016 NASA and ESA used the Hubble telescope to measure the distances to stars in nineteen galaxies. The results, obtained by a team under Adam Riess, used high-precision statistical methods to revise the Hubble constant upwards to 73·2 kilometres per second per megaparsec.[1] This means that the universe is expanding 5–9% faster than previously thought. With the standard model of cosmology, this figure no longer agrees with observations of the cosmic microwave background by WMAP and ESA's Planck satellite. This unexpected result may be a new clue to the nature of dark matter and dark energy, or a sign that neither exists and that our picture of the universe needs revising.

That's how real science advances, of course. Three steps forward, two steps back. Mathematicians have the luxury of living in a logical bubble, where once something is proved true, it *remains* true. Interpretations and proofs may change, but the theorems don't get unproved by later discoveries. Though they may become obsolete or irrelevant to current concerns. Science is always provisional, only as good as the current evidence. In response to such evidence, scientists reserve the right to *change their minds*.

Even when we think we understand something, unexpected issues arise. Theoretically, all sorts of variations on our universe make just as much sense as this one. When calculations seemed to indicate that most of the variants could not support life, or even atoms, the philosophical puzzle of fine-tuning made its grand entrance onstage. Attempts to solve it led to some of the most imaginative, though speculative, ideas that physicists have yet devised. However, none of them is necessary if, as a closer analysis of the reasoning suggests, the whole problem is a red herring.

The main thrust of *Calculating the Cosmos* is the need for, and the astonishing success of, mathematical reasoning in astronomy and cosmology. Even when I've criticised popular theories, I've started by explaining the conventional view, and why so many people go along with it. But when there seem to be good reasons to consider alternatives, and especially when those alternatives aren't being taken seriously, I think it's worth presenting them – even if they're controversial or rejected by many cosmologists. I don't want you to accept confident claims to have solved the riddles of the universe when many unresolved issues remain. On the other hand, I want to explain the conventional solutions too: they're beautiful applications of mathematics, they may well be right, and if not, they're paving the way towards something better.

The alternatives often seem radical – there was no Big Bang, dark matter is a chimera, ... But only a few decades ago, neither theory had any proponents at all. Research at the distant frontiers of knowledge is always difficult, and we can't install the universe in a laboratory, put it under a microscope, distil it to find out what it's made of, or stress it to see what breaks. We have to employ inference and imagination. Along with our critical faculties, which is why I've placed more emphasis than is common on ideas that don't reflect the conventional wisdom. Those are valid parts of the scientific process too.

We've come across dozens of erroneous theories that seemed entirely reasonable not so long ago. The Earth is the centre of the universe. The planets formed when a passing star pulled out a cigar-shaped mass from the Sun. There's a planet orbiting closer to the Sun than Mercury. Saturn has ears. The Sun is the only star with planets. The Galaxy is at rest at the centre of the universe, surrounded by an infinite void. The distribution of galaxies is smooth. The universe has always existed, but new matter is created in the interstellar void. In their time these theories were widely believed, and most were based on the best evidence then available. Some were a bit silly all along, mind you; scientists sometimes get very strange ideas, reinforced by herd instinct and a quasi-religious fervour rather than evidence.

I see no reason why today's treasured theories should fare any better. Maybe the Moon wasn't created by the Earth colliding with a body the size of Mars. Maybe there was no Big Bang. Maybe redshift isn't evidence for an expanding universe. Maybe black holes don't exist. Maybe inflation never happened. Maybe dark matter is a mistake. Maybe alien life can be radically different from anything we've ever encountered, possibly even than we can imagine.

Maybe.

Maybe not.

The fun will be finding out.

Units and Jargon

antimatter Matter composed of antiparticles, having the same masses as ordinary particles but opposite charges.

asteroid Small rocky or icy body orbiting the Sun, mainly between Mars and Jupiter.

astronomical unit (AU) Distance from the Sun to the Earth, 149,597,871 kilometres.

Big Bang Theory that the universe originated in a singularity 13·8 billion years ago.

black-body radiation Spectrum of electromagnetic radiation from an opaque non-reflective body at constant temperature.

black hole Region of space from which light can't escape, often formed by a massive star collapsing under its own gravity.

centaur Body occupying an orbit that crosses the ecliptic between the orbits of Jupiter and Neptune.

comet Small icy body that heats up when it gets close to the Sun, showing a visible atmosphere (coma) and perhaps a tail, caused by gas streaming away on the solar wind.

cosmic microwave background (CMB) Almost uniform radiation at a temperature of 3 K generally believed to be a relic of the Big Bang.

curvature Intrinsic measure of how a surface or manifold differs from flat Euclidean space.

degree (Angles) 360 degrees equal a full circle.

(Temperature) The units used in this book are Celsius (°C) and kelvin (K). Celsius runs from 0 (water freezes) to 100 (water turns to steam). Kelvin is Celsius plus 273·16, and 0 K (−273·16°C) is the lowest possible temperature, absolute zero.

Dermott's law The orbital period of a planet's nth satellite is proportional to the nth power C^n of a constant C. Different satellite systems can have different constants.

eccentricity Measure of how thin or fat an ellipse is. See Note 2 of Chapter 1.

electronvolt (eV) Unit of energy used in particle physics, equal to $1 \cdot 6 \times 10^{-19}$ joules. See joule.

ellipse Closed oval curve formed by stretching a circle uniformly in one direction.

event horizon Boundary of black hole, through which light can't escape.

exomoon Natural satellite of a planet orbiting a star other than the Sun.

exoplanet Planet orbiting a star other than the Sun.

fine structure constant α Fundamental constant characterising the strength of interaction between charged particles. Equal to 7.297352×10^{-3}. (It is a dimensionless number, independent of units of measurement.)

gamma ray Form of electromagnetic radiation consisting of high-energy photons.

gamma-ray burster Source of sudden bursts of gamma rays, thought to be of two kinds: the formation of a neutron star or a black hole, or the merger of a binary pair of neutron stars.

GeV (gigaelectronvolt) Unit of energy used in particle physics. One billion electronvolts. See electronvolt.

gravitational constant G Constant of proportionality in Newton's law of gravity. Equal to 6.674080×10^{-11} m^3 kg^{-1} s^{-2}.

isotope Variant of a chemical element, distinguished by how many neutrons its atom has.

joule (J) Unit of energy, producing one watt of power for one second. (A single bar of an electric fire typically produces 1000 watts.)

Kelvin (K) See degree.

Kepler's laws of planetary motion

1   The orbit of a planet is an ellipse with the Sun at one focus.
2   The line from the Sun to the planet sweeps out equal areas in equal periods of time.

3 The square of the period of revolution is proportional to the cube of the distance.

Liapunov time Timescale on which a dynamical system is chaotic. The time for the distance between nearby trajectories to increase by a factor of e ~ 2·718. Sometimes e is replaced by 2 or 10. Related to the prediction horizon, beyond which forecasts become unreliable.

light cone Region of spacetime accessible from a given event by following a timelike curve or worldline.

light curve How a star's output of radiation varies with time.

lightspeed See speed of light.

light year Distance light travels in a year, $9·460528 \times 10^{15}$ metres, or 9·46 trillion kilometres.

luminosity Total energy radiated by a star per unit of time. Measured in joules per second (watts). The Sun's luminosity is $3·846 \times 10^{26}$ watts.

magnitude Logarithmic measure of brightness. Apparent magnitude is the brightness as seen from Earth, absolute magnitude is how bright it would be seen from a distance of 10 parsecs (for stars) and 1 astronomical unit (for asteroids and planets). Brighter objects have lower magnitudes, which can be negative. The Sun has apparent magnitude −27, the full Moon −13, Venus −5, and the brightest visible star Sirius −1·5. A decrease in magnitude of 5 corresponds to a 100-fold increase in brightness.

major axis Longest axis of an ellipse.

major radius Half the length of longest axis of an ellipse.

manifold Multidimensional smooth space, like a surface but with any number of coordinates.

MeV (megaelectronvolt) Unit of energy used in particle physics. One million electronvolts. See electronvolt.

minor axis Shortest axis of an ellipse.

minor radius Half the length of shortest axis of an ellipse.

minute (of arc) 60 minutes = 1 degree.

Newton's law of gravity Every body attracts every other body with a force proportional to their masses, and inversely proportional to the square of the distance between them. The constant of proportionality is called the gravitational constant.

Newton's laws of motion

1 Bodies continue to move in a straight line at a constant speed unless acted on by a force.

2 The acceleration of any body, multiplied by its mass, is equal to the force acting on it.

3 Every action produces an equal and opposite reaction.

oblate Flattened at the poles.

occultation When one celestial object appears to pass behind another, which hides it. Especially used for stars occulted by a moon or planet.

parallax Half the difference in the directions to a star from opposite points on the Earth's orbit, on a diameter at right angles to the line from the Sun to the star.

parsec Distance to a star with parallax one second of arc, 3·26 light years.

perihelion Point of closest approach to the Sun.

period Time in which periodically recurring behaviour repeats. Examples are the period of revolution of a planet about its primary (roughly 365 days for Earth), or the period of rotation (24 hours for Earth).

Planck's constant h A basic constant in quantum mechanics, determining the minimal amount of energy in an electromagnetic wave (which is h times the frequency). It is very small, equal to $1·054571 \times 10^{-34}$ joule-seconds.

planetesimal Small bodies that can aggregate to form planets, thought to have been common in the early solar system.

precession Slow rotation of the axis of an elliptical orbit.

primary Parent body round which the body concerned orbits. The Earth's primary is the Sun, the Moon's primary is the Earth.

resonance Coincidence of timing in which the periods of two repeating effects are in a simple fractional relationship. See Note 6 of Chapter 2.

satellite (Natural) Smaller body in orbit round a planet, a 'moon'. (Artificial) human-made machine in orbit round the Earth or some other body of the solar system.

second (of arc) 60 seconds = 1 minute.

spacetime Four-dimensional manifold with three space coordinates and one time coordinate.

spectrum How the amount of radiation emitted by a body (usually a star) varies with wavelength. Peaks (emission lines) and dips (absorption lines) are the most important features.

speed of light c Equal to 299,792,458 metres per second.

spin–orbit resonance Fractional relationship between a body's period of rotation about its axis and its period of revolution about its primary.

Titius–Bode law The distance from the Sun to the nth planet is $0.075 \times 2^n + 0.4$ astronomical units.

torus Mathematical surface shaped like an American doughnut (with a hole).

Trans-Neptunian object (TNO) Asteroid or other small body orbiting the Sun at an average distance greater than that of Neptune (30 AU).

Notes and References

Prologue

1. Mars Odyssey, Mars Express, MRO, Mars Orbiter Mission, and MAVEN.

2. NASA's *Opportunity* and *Curiosity* rovers. The *Spirit* rover ceased functioning in 2011.

3. 'This foolish idea of shooting at the Moon is an example of the absurd length to which vicious specialisation will carry scientists. To escape the Earth's gravitation a projectile needs a velocity of 7 miles per second. The thermal energy at this speed is 15,180 calories [per gram]. Hence the proposition appears to be basically impossible.' Alexander Bickerton, Chemistry Professor, 1926.

    'I am bold enough to say that a man-made Moon voyage will never occur regardless of all scientific advances.' Lee De Forest, electronics inventor, 1957.

    'There is no hope for the fanciful idea of reaching the Moon because of insurmountable barriers to escaping the Earth's gravity.' Forest Moulton, astronomer, 1932.

4. In a 1920 editorial, the *New York Times* wrote: 'Professor Goddard … does not know the relation of action to re-action, and the need to have something better than a vacuum against which to react.' Newton's third law of motion states that to every action there is an equal and opposite reaction. The reaction comes from conservation of momentum, and no medium to react *against* is required. Such a medium would impede progress, not assist it. To be fair, the newspaper apologised in 1969 when the *Apollo 11* astronauts were on their way to the Moon. To every publication there is an equal and opposite retraction.

5. Nicolas Bourbaki is the pseudonym of an ever-changing group of mainly French mathematicians first formed in 1935, who wrote a long series of books reformulating mathematics on a general and

abstract basis. This was great for research mathematics, because it unified the subject, sorted out basic concepts, and provided rigorous proofs. But the widespread adoption of a similar philosophy in the teaching of school mathematics, known as 'new maths' met with little success, and was, to say the least, controversial.

1. Attraction at a Distance

1. In 1726 Newton spent an evening dining with William Stukeley in London. In a document preserved in the archives of the Royal Society, whose archaic spelling I've preserved for period flavour, Stukeley wrote:

   'After dinner, the weather being warm, we went into the garden & drank thea under the shade of some apple tree; only he & myself. Amid other discourse, he told me, he was just in the same situation, as when formerly the notion of gravitation came into his mind. Why shd that apple always descend perpendicularly to the ground, thought he to himself; occasion'd by the fall of an apple, as he sat in contemplative mood. Why shd it not go sideways, or upwards? But constantly to the Earth's centre? Assuredly the reason is, that the Earth draws it. There must be a drawing power in matter. And the sum of the drawing power in the matter of the Earth must be in the Earth's centre, not in any side of the Earth. Therefore does this apple fall perpendicularly or towards the centre? If matter thus draws matter; it must be in proportion of its quantity. Therefore the apple draws the Earth, as well as the Earth draws the apple.'

   Other sources also confirm that Newton told this story, but none of this proves the story true. Newton might have invented it to explain his ideas. A still extant tree – Flower of Kent, a cooking apple, at Woolsthorpe Manor – is said to be the one from which the apple fell.

2. If an ellipse has major radius a and minor radius b, then the focus lies at a distance $f = \sqrt{a^2 - b^2}$ from the centre. The eccentricity is $\varepsilon = f / a \sqrt{1 - b^2 / a^2}$.

3. A. Koyré. An unpublished letter of Robert Hooke to Isaac Newton, *Isis* **43** (1952) 312–337.

4. A. Chenciner and R. Montgomery. A remarkable periodic solution of the three-body problem in the case of equal masses, *Ann. Math.* **152** (2000) 881–901.

An animation, and further information about similar types of orbit, is at: http://www.scholarpedia.org/article/N-body_choreographies

5. C. Simó. New families of solutions in *N*-body problems, *Proc. European Congr. Math.*, Barcelona, 2000.

6. E. Oks. Stable conic-helical orbits of planets around binary stars: analytical results, *Astrophys. J.* **804** (2015) 106.

7. Newton put it this way in a letter to Richard Bentley, written in 1692 or 1693: 'It is inconceivable that inanimate Matter should, without the Mediation of something else, which is not material, operate upon, and affect other matter without mutual Contact… That one body may act upon another at a distance thro' a Vacuum, without the Mediation of any thing else … is to me so great an Absurdity that I believe no Man who has in philosophical Matters a competent Faculty of thinking can ever fall into it.'

8. That's slightly simplistic. Passing *through* lightspeed is what's forbidden. Nothing currently moving slower than light can speed up to become faster than light; if anything happens to be moving faster than light it can't decelerate to become slower than light. Particles like this are called tachyons: they're entirely hypothetical.

9. In a letter of 1907 to his friend Conrad Habicht, Einstein wrote that he was thinking about 'a relativistic theory of the gravitational law with which I hope to account for the still unexplained secular change in the perihelion motion of Mercury'. His first significant attempts began in 1911.

10. Nowadays we combine Einstein's equations into a single tensor equation (with ten components – a symmetric 4-tensor). But 'field equations' remains the standard name.

2. Collapse of the Solar Nebula

1. The oldest minerals found in meteorites, modern traces of the first solid material in the pre-solar nebula, are 4·5682 billion years old.

2. He wrote it in 1662–3, but postponed publication because of the Inquisition. It appeared shortly after his death.

3. A proper definition requires vectors.

4. H. Levison, K. Kretke, and M. Duncan. Growing the gas-giant planets by the gradual accumulation of pebbles, *Nature* **524** (2015) 322–324.

5. I. Stewart. The second law of gravitics and the fourth law of thermodynamics, in *From Complexity to Life* (ed. N.H. Gregsen), Oxford University Press, 2003, pp. 114–150.

6. In this book the notation $p:q$ for a resonance means that the first body mentioned goes round p times while the second goes round q times. Their *periods* are therefore in the ratio q/p. On the other hand their *frequencies* are in the ratio p/q. Some authors use the opposite convention; others use the notation 'p/q resonance'. Reversing the order of the bodies turns a $p:q$ resonance into a $q:p$ resonance.

7. Venus doesn't have old craters because its surface was reshaped by volcanism less than a hundred million years ago. The planets from Jupiter outwards are gas and ice giants, and all we can see is their upper atmosphere. But many of their moons have craters – some new, some old. *New Horizons* revealed that Pluto and its moon Charon have fewer craters than expected.

8. K. Batygin and G. Laughlin. On the dynamical stability of the solar system, *Astrophys. J.* **683** (2008) 1207–1216.

9. J. Laskar and M. Gastineau. Existence of collisional trajectories of Mercury, Mars and Venus with the Earth, *Nature* **459** (2009) 817–819.

10. G. Laughlin. Planetary science: The Solar System's extended shelf life, *Nature* **459** (2009) 781–782.

3. Inconstant Moon

1. The chemistry of uranium deposits at Oklo, Gabon, suggests that in the Precambrian they constituted a natural fission reactor.

2. R.C. Paniello, J.M.D. Day, and F. Moynier. Zinc isotopic evidence for the origin of the Moon, *Nature* **490** (2012) 376–379.

3. A.G.W. Cameron and W.R. Ward. The origin of the Moon, *Abstr. Lunar Planet. Sci. Conf.* **7** (1976) 120–122.

4. W. Benz, W.L. Slattery, and A.G.W. Cameron. The origin of the moon and the single impact hypothesis I, *Icarus* **66** (1986) 515–535.

W. Benz, W.L. Slattery, and A.G.W. Cameron. The origin of the moon and the single impact hypothesis II, *Icarus* **71** (1987) 30–45.

W. Benz, A.G.W. Cameron, and H.J. Melosh. The origin of the moon and the single impact hypothesis III, *Icarus* **81** (1989) 113–131.

5. R.M. Canup and E. Asphaug. Origin of the Moon in a giant impact near the end of the Earth's formation, *Nature* **412** (2001) 708–712.

6. A. Reufer, M.M.M. Meier, and W. Benz. A hit-and-run giant impact scenario, *Icarus* **221** (2012) 296–299.

7. J. Zhang, N. Dauphas, A.M. Davis, I. Leya, and A. Fedkin. The proto-Earth as a significant source of lunar material, *Nature Geosci.* **5** (2012) 251–255.

8. R.M. Canup, Simulations of a late lunar-forming impact, *Icarus* **168** (2004) 433–456.

9. A. Mastrobuono-Battisti, H.B. Perets, and S.N. Raymond. A primordial origin for the compositional similarity between the Earth and the Moon, *Nature* **520** (2015) 212–215.

4. The Clockwork Cosmos

1. See note 6 of Chapter 2 for why we don't call it a 3:5 resonance.

2. Dermott's law, an empirical formula for the orbital period of satellites in the solar system, was identified by Stanley Dermott in the 1960s. It takes the form $T(n) = T(0)C^n$, where $n = 1, 2, 3, 4, \dots$. Here $T(n)$ is the orbital period of the nth satellite, $T(0)$ is a constant of the order of days, and C is a constant of the satellite system in question. Specific values are: *Jupiter*: $T(0) = 0.444$ days, $C = 2.0$. *Saturn*: $T(0) = 0.462$ days, $C = 1.59$. *Uranus*: $T(0) = 0.488$ days, $C = 2.24$.

S.F. Dermott. On the origin of commensurabilities in the solar system II: the orbital period relation, *Mon. Not. RAS* **141** (1968) 363–376.

S.F. Dermott. On the origin of commensurabilities in the solar system III: the resonant structure of the solar system, *Mon. Not. RAS* **142** (1969) 143–149.

3. F. Graner and B. Dubrulle. Titius-Bode laws in the solar system. Part I: Scale invariance explains everything, *Astron. & Astrophys.* **282** (1994) 262–268.

B. Dubrulle and F. Graner. Titius-Bode laws in the solar system. Part II: Build your own law from disk models, *Astron. & Astrophys.* **282** (1994) 269–276.

4. Derived from 'QB1-o', after (15760) 1992 QB₁, the first TNO discovered.

5. It's tricky to measure the diameter of Pluto from Earth, even using the Hubble telescope, because it has a thin atmosphere that makes its edges fuzzy. Eris has no atmosphere.

6. Propositions 43–45 of Book I of *Philosophiae Naturalis Principia Mathematica*.

7. A.J. Steffl, N.J. Cunningham, A.B. Shinn, and S.A. Stern. A search for Vulcanoids with the STEREO heliospheric imager, *Icarus* **233** (2013) 48–56.

5. Celestial Police

1. Wigner's remark is often misunderstood. It's easy to explain the *effectiveness* of mathematics. Much of it is motivated by real-world problems, so it's no surprise when it solves those problems. The important word in Wigner's phrase is 'unreasonable'. He was referring to the way mathematics invented for one purpose often turns out to be useful in a totally different, unexpected area. Simple examples are Greek geometry of conic sections, turning up in planetary orbits two thousand years later, or Renaissance speculations about imaginary numbers, now central to mathematical physics and engineering. This widespread phenomenon can't be explained away so easily.

2. Suppose, for simplicity, that all asteroids lie in the same plane – which isn't too far from reality for most. The asteroid belt lies between 2·2 and 3·3 AU from the Sun, that is, about 320 million and 480 million kilometres. Projected into the plane of the ecliptic, the total area occupied by the asteroid belt is $\pi(480^2 - 320^2)$ *trillion* square kilometres, that is, 4×10^{17} km². Shared among 150 million rocks this gives an area of $8·2 \times 10^8$ km² per rock. That's the same area as a circle of diameter 58,000 km. If the asteroids are roughly uniformly distributed, which is good enough for government work, that's the typical distance between neighbouring asteroids.

3. M. Moons and A. Morbidelli. Secular resonances inside mean-motion commensurabilities: the 4/1, 3/1, 5/2 and 7/3 cases, *Icarus* **114** (1995) 33–50.

    M. Moons, A. Morbidelli, and F. Migliorini. Dynamical structure of the 2/1 commensurability with Jupiter and the origin of the resonant asteroids, *Icarus* **135** (1998) 458–468.

4. An animation showing the relationship between the five Lagrangian points and the gravitational potential is at https://en.wikipedia.org/wiki/File:Lagrangian_points_equipotential.gif

5. See the animation at https://www.exploremars.org/trojan-asteroids-around-jupiter-explained.

6. F.A. Franklin. Hilda asteroids as possible probes of Jovian migration, *Astron. J.* **128** (2004) 1391–1406.

7. http://www.solstation.com/stars/jupiter.htm

6. The Planet that Swallowed its Children

1. P. Goldreich and S. Tremaine. Towards a theory for the Uranian rings, *Nature* **277** (1979) 97–99.

2. M. Kenworthy and E. Mamajek. Modeling giant extrasolar ring systems in eclipse and the case of J1407b: sculpting by exomoons? arXiv:1501.05652 (2015).

3. F. Braga-Rivas and 63 others. A ring system detected around Centaur (10199) Chariklo, *Nature* **508** (2014) 72–75.

7. Cosimo's Stars

1. E.J. Rivera, G. Laughlin, R.P. Butler, S.S. Vogt, N. Haghighipour, and S. Meschiari. The Lick-Carnegie exoplanet survey: a Uranus-mass fourth planet for GJ 876 in an extrasolar Laplace configuration, *Astrophys. J.* **719** (2010) 890–899.

2. B.E. Schmidt, D.D. Blankenship, G.W. Patterson, and P.M. Schenk. Active formation of 'chaos terrain' over shallow subsurface water on Europa, *Nature* **479** (2011) 502–505.

3. P.C. Thomas, R. Tajeddine, M.S. Tiscareno, J.A. Burns, J. Joseph, T.J. Loredo, P. Helfenstein, and C. Porco. Enceladus's measured

physical libration requires a global subsurface ocean, *Icarus* (2015) in press; doi:10.1016/j.icarus.2015.08.037.

4. S. Charnoz, J. Salmon, and A. Crida. The recent formation of Saturn's moonlets from viscous spreading of the main rings, *Nature* **465** (2010) 752–754.

8. Off on a Comet

1. M. Massironi and 58 others. Two independent and primitive envelopes of the bilobate nucleus of comet 67P, *Nature* **526** (2015) 402–405.

2. A. Bieler and 33 others. Abundant molecular oxygen in the coma of comet 67P/Churyumov–Gerasimenko, *Nature* **526** (2015) 678–681.

3. P. Ward and D. Brownlee. *Rare Earth*, Springer, New York, 2000.

4. J. Horner and B.W. Jones. Jupiter – friend or foe? I: The asteroids, *Int. J. Astrobiol.* **7** (2008) 251–261.

9. Chaos in the Cosmos

1. See the video at http://hubblesite.org/newscenter/archive/releases/2015/24/video/a/

2. J.R. Buchler, T. Serre, and Z. Kolláth. A chaotic pulsating star: the case of R Scuti, *Phys. Rev. Lett.* **73** (1995) 842–845.

3. Strictly, 'dice' is the plural and 'die' is the singular, but in language as she is actually spoke, almost everyone talks of 'a dice'. I no longer see any point in fighting this, but I'm not using the word out of ignorance. I'm still fighting a rearguard action on usages such as 'the team are', but deep down I know I've lost that one too. I've also stopped trying to tell greengrocers the difference between the plural and the possessive, though I was sorely tempted to have a quiet chat with the bloke up the road whose van bears the sign: REMOVAL'S.

4. Nonetheless, a 6 has the same probability as any other value, for a fair dice. In the long run, the numbers of 6s should get arbitrarily close to 1/6 of the number of throws. But how this happens is instructive. If at some stage there have been, say, 100 more throws of a 6 than anything else, a 6 doesn't become more likely. The dice just keeps churning out more and more numbers. After, say, a hundred million more throws, that extra 100 affects the proportion of 6s by

only one part in a million. Deviations aren't cancelled out because the dice 'knows' it's thrown too many 6s. They're diluted by new data, generated by a dice that has no memory.

5. Dynamically, a dice is a solid cube, and its motion is chaotic because the edges and corners 'stretch' the dynamics. But there's another source of randomness in dice: initial conditions. How you hold the dice in your hand, and how you release it, randomise the result anyway.

6. Lorenz didn't call it a butterfly, though he did say something similar about a seagull. Someone else came up with the butterfly for the title of a public lecture Lorenz gave in 1972. And what Lorenz originally had in mind probably wasn't *this* butterfly effect, but a subtler one. See: T. Palmer. The real butterfly effect, *Nonlinearity* 27 (2014) R123–R141.

   None of that affects this discussion, and what I've described is what we now mean by 'butterfly effect'. It's real, it's characteristic of chaos, but it's subtle.

7. V. Hoffmann, S.L. Grimm, B. Moore, and J. Stadel. Chaos in terrestrial planet formation, *Mon. Not. RAS* (2015); arXiv: 1508.00917.

8. A. Milani and P. Farinella. The age of the Veritas asteroid family deduced by chaotic chronology, *Nature* 370 (1994) 40–42.

9. June Barrow-Green. *Poincaré and the Three Body Problem*, American Mathematical Society, Providence, 1997.

10. M.R. Showalter and D.P. Hamilton. Resonant interactions and chaotic rotation of Pluto's small moons, *Nature* 522 (2015) 45–49.

11. J. Wisdom, S.J. Peale, and F. Mignard. The chaotic rotation of Hyperion, *Icarus* 58 (1984) 137–152.

12. K = *Kreide*, German for 'chalk', referring to the Cretaceous, and T = Tertiary. Why do scientists do this kind of thing? Beats me.

13. M.A. Richards and nine others. Triggering of the largest Deccan eruptions by the Chicxulub impact, *GSA Bull.* (2015), doi: 10.1130/B31167.1.

14. W.F. Bottke, D. Vokrouhlický, and D. Nesvorný . An asteroid breakup 160 Myr ago as the probable source of the K/T impactor, *Nature* 449 (2007) 48–53.

10. The Interplanetary Superhighway

1. M. Minovitch. A method for determining interplanetary free-fall reconnaissance trajectories, *JPL Tech. Memo.* TM-312–130 (1961) 38–44.

2. M. Lo and S. Ross. SURFing the solar system: invariant manifolds and the dynamics of the solar system, *JPL IOM* 312/97, 1997.

   M. Lo and S. Ross. The Lunar L1 gateway: portal to the stars and beyond, *AIAA Space 2001 Conf.*, Albuquerque, 2001.

3. http://sci.esa.int/where_is_rosetta/ has a dramatic animation of this roundabout path.

4. One cause (among many) of World War I was the assassination of the Austrian Archduke Franz Ferdinand on a visit to Sarajevo. Six assassins made a failed attempt with a grenade. Later one of them, Gavrilo Princip, shot him dead with a pistol, along with his wife Sophie. Initial reaction by the populace was virtually non-existent, but the Austrian government encouraged rioting against Serbs in Sarajevo, which escalated.

5. W.S. Koon, M.W. Lo, J.E. Marsden, and S.D. Ross. The Genesis trajectory and heteroclinic connections, *Astrodynamics* 103 (1999) 2327–2343.

11. Great Balls of Fire

1. Strictly speaking, this term refers to total energy output, but that's closely related to intrinsic brightness.

2. An animation of stellar evolution across the Hertzsprung–Russell diagram can be found at http://spiff.rit.edu/classes/phys230/lectures/star_age/evol_hr.swf

3. F. Hoyle. Synthesis of the elements from hydrogen, *Mon. Not. RAS* 106 (1946) 343–383.

4. E.M. Burbidge, G.R. Burbidge, W.A. Fowler, and F. Hoyle. Synthesis of the elements in stars, *Rev. Mod. Phys.* 29 (1957) 547–650.

5. A.J. Korn, F. Grundahl, O. Richard, P.S. Barklem, L. Mashonkina, R. Collet, N. Piskunov, and B. Gustafsson. A probable stellar solution to the cosmological lithium discrepancy, *Nature* 442 (2006) 657–659.

6. F. Hoyle. On nuclear reactions occurring in very hot stars: the synthesis of the elements between carbon and nickel, *Astrophys. J. Suppl.* **1** (1954) 121–146.

7. F. Hoyle. The universe: past and present reflections, *Eng. & Sci.* (November 1981) 8–12.

8. G.H. Miller and 12 others. Abrupt onset of the Little Ice Age triggered by volcanism and sustained by sea-ice/ocean feedbacks, *Geophys. Res. Lett.* **39** (2012) L02708.

9. H.W. Babcock. The topology of the Sun's magnetic field and the 22-year cycle, *Astrophys. J.* **133** (1961) 572–587.

10. E. Nesme-Ribes, S.L. Baliunas, and D. Sokoloff. The stellar dynamo, *Scientific American* (August 1996) 30–36.

    For mathematical details and more recent work with more realistic models, see: M. Proctor. Dynamo action and the Sun, *EAS Publ. Ser.* **21** (2006) 241–273.

12. Great Sky River

1. That is, $M(r) = rv(r)^2/G$. So $v(r) = \sqrt{GM(r)/r}$. Here $M(r)$ is the mass out to radius r, $v(r)$ is the rotational velocity of stars at radius r, and G is the gravitational constant.

13. Alien Worlds

1. X. Dumusque and 10 others. An Earth-mass planet orbiting α Centauri B, *Nature* **491** (2012) 207–211.

2. V. Rajpaul, S. Aigrain, and S.J. Roberts. Ghost in the time series: no planet for Alpha Cen B, arXiv:1510.05598; *Mon. Not. RAS*, in press.

3. Z.K. Berta-Thompson and 20 others. A rocky planet transiting a nearby low-mass star, *Nature* **527** (2015) 204–207.

4. 'Earthlike' here means a rocky world, with much the same size and mass as the Earth, in an orbit that would allow water to exist as a liquid without any special extra conditions. Later we require oxygen as well.

5. E. Thommes, S. Matsumura, and F. Rasio. Gas disks to gas giants: Simulating the birth of planetary systems, *Nature* **321** (2008) 814–817.

6. M. Hippke and D. Angerhausen. A statistical search for a population of exo-Trojans in the Kepler dataset, ArXiv:1508.00427 (2015).

7. In *Evolving the Alien* Cohen and I propose that what really counts is *extelligence*: the ability of intelligent beings to pool their knowledge in a way that all can access. The Internet is an example. It takes extelligence to build starships.

8. M. Lachmann, M.E.J. Newman, and C. Moore. The physical limits of communication, Working paper 99–07–054, Santa Fe Institute 2000.

9. I.N. Stewart. Uninhabitable zone, *Nature* 524 (2015) 26.

10. P.S. Behroozi and M. Peeples. On the history and future of cosmic planet formation, *Mon. Not. RAS* (2015); arXiv: 1508.01202.

11. D. Sasselov and D. Valencia. Planets we could call home, *Scientific American* 303 (August 2010) 38–45.

12. S.A. Benner, A. Ricardo, and M.A. Carrigan. Is there a common chemical model for life in the universe? *Current Opinion in Chemical Biology* 8 (2004) 676–680.

13. J. Stevenson, J. Lunine, and P. Clancy. Membrane alternatives in worlds without oxygen: Creation of an azotosome, *Science Advances* 1 (2015) e1400067.

14. J. Cohen and I. Stewart. *Evolving the Alien*, Ebury Press, London, 2002.

15. W. Bains. Many chemistries could be used to build living systems, *Astrobiology* 4 (2004) 137–167.

16. J. von Neumann. *Theory of Self-Reproducing Automata*, University of Illinois Press, Urbana, 1966.

14. Dark Stars

1. In units that make the speed of light equal to 1, say years for time and light years for space.

2. R. Penrose. Conformal treatment of infinity, in *Relativity, Groups and Topology* (ed. C. de Witt and B. de Witt), Gordon and Breach, New York, 1964, pp. 563–584; *Gen. Rel. Grav.* 43 (2011) 901–922.

3. Animations of what it would look like when passing through these wormholes can be found at http://jila.colorado.edu/~ajsh/insidebh/penrose.html

4. B.L. Webster and P. Murdin. Cygnus X-1 – a spectroscopic binary with a heavy companion?, *Nature* **235** (1972) 37–38.

    H.L. Shipman, Z. Yu, and Y.W. Du. The implausible history of triple star models for Cygnus X-1: Evidence for a black hole, *Astrophys. Lett.* **16** (1975) 9–12.

5. P. Mazur and E. Mottola. Gravitational condensate stars: An alternative to black holes, arXiv:gr-qc/0109035 (2001).

15. Skeins and Voids

1. Colin Stuart. When worlds collide, *New Scientist* (24 October 2015) 30–33.

2. You may object that 'currently' has no meaning because relativity implies that events need not occur simultaneously for all observers. That's true, but when I say 'currently' I'm referring to *my* frame of reference, with me as observer. I can conceptually set distant clocks by making changes of one year per light year; viewed from here, they will all be synchronised. More generally, observers in 'comoving' frames experience simultaneity the way we would expect in classical physics.

3. N.J. Cornish, D.N. Spergel, and G.D. Starkman. Circles in the sky: finding topology with the microwave background radiation, *Classical and Quantum Gravity* **15** (1998) 2657–2670.

    J.R. Weeks. Reconstructing the global topology of the universe from the cosmic microwave background, *Classical and Quantum Gravity* **15** (1998) 2599–2604.

16. The Cosmic Egg

1. Less than that! According to NASA it was 12% of a pixel.

2. Based on Type Ia supernovae, temperature fluctuations in the CMB, and the correlation function of galaxies, the universe has an estimated age of 13.798 ± 0.037 billion years. See Planck collaboration (numerous authors). Planck 2013 results XVI:

Cosmological parameters, *Astron. & Astrophys.* **571** (2014); arXiv:1303.5076.

3. M. Alcubierre. The warp drive: hyper-fast travel within general relativity, *Classical and Quantum Gravity* **11** (1994). L73–L77.

    S. Krasnikov. The quantum inequalities do not forbid spacetime shortcuts, *Phys. Rev. D* **67** (2003) 104013.

4. See Note 2 of Chapter 15 on simultaneity in a relativistic universe.

17. The Big Blow-Up

1. The current figure for the temperature is $2·72548 \pm 0·00057$ K, see D.J. Fixsen. The temperature of the cosmic microwave background, *Astrophys. J.* **707** (2009) 916–920.

    Other figures mentioned in the text are historical estimates, now obsolete.

2. This phrase is reused from Terry Pratchett, Ian Stewart, and Jack Cohen. *The Science of Discworld IV: Judgement Day*, Ebury, London, 2013.

3. Penrose's work is reported in: Paul Davies. *The Mind of God*, Simon & Schuster, New York, 1992.

4. G.F.R. Ellis. Patchy solutions, *Nature* **452** (2008) 158–161.

    G.F.R. Ellis. The universe seen at different scales, *Phys. Lett.* A **347** (2005) 38–46.

5. T. Buchert. Dark energy from structure: a status report, *T. Gen. Rel. Grav.* **40** (2008) 467–527.

6. J. Smoller and B. Temple. A one parameter family of expanding wave solutions of the Einstein equations that induces an anomalous acceleration into the standard model of cosmology, arXiv:0901.1639.

7. R.R. Caldwell. A gravitational puzzle, *Phil. Trans. R. Soc. London* A **369** (2011) 4998–5002.

8. R. Durrer. What do we really know about dark energy? *Phil. Trans. R. Soc. London* A **369** (2011) 5102–5114.

9. Marcus Chown. End of the beginning, *New Scientist* (2 July 2005) 30–35.

10. D.J. Fixsen. The temperature of the cosmic microwave background, *Astrophys. J.* **707** (2009) 916–920.

11. The stars in galaxies are bound together by gravity, which is thought to counteract the expansion.

12. S. Das, Quantum Raychaudhuri equation, *Phys. Rev.* D **89** (2014) 084068.

    A.F. Ali and S. Das. Cosmology from quantum potential, *Phys. Lett.* B **741** (2015) 276–279.

13. Jan Conrad. Don't cry wolf, *Nature* **523** (2015) 27–28.

18. The Dark Side

1. Quasi-autonomous non-governmental organisation.

2. K.N. Abazajian and E. Keeley. A bright gamma-ray galactic center excess and dark dwarfs: strong tension for dark matter annihilation despite Milky Way halo profile and diffuse emission uncertainties, arXiv: 1510.06424 (2015).

3. G. R. Ruchti and 28 others. The Gaia-ESO Survey: a quiescent Milky Way with no significant dark/stellar accreted disc, *Mon. Not. RAS* **450** (2015) 2874–2887.

4. S. Clark. Mystery of the missing matter, *New Scientist* (23 April 2011) 32–35.

    G. Bertone, D. Hooper, and J. Silk. Particle dark matter: evidence, candidates and constraints, *Phys. Rep.* **405** (2005) 279–390.

5. Newton's second law of motion is $F = ma$, where F = force, m = mass, a = acceleration. MOND replaces this by $F = \mu(a/a_0)ma$, where a_0 is a new fundamental constant that determines the acceleration below which Newton's law ceases to apply. The term $\mu(x)$ is an unspecified function that tends to 1 as x becomes large, in agreement with Newton's law, but to x when x is small, which models observed galactic rotation curves.

6. J.D. Bekenstein, Relativistic gravitation theory for the modified Newtonian dynamics paradigm, *Physical Review* D **70** (2004) 083509.

7. D. Clowe, M. Bradač, A.H. Gonzalez, M. Markevitch, S.W. Randall, C. Jones, and D. Zaritsky. A direct empirical proof of the existence of dark matter, *Astrophys. J. Lett.* **648** (2006) L109.

8. http://www.astro.umd.edu/~ssm/mond/moti_bullet.html

9. S. Clark. Mystery of the missing matter, *New Scientist* (23 April 2011) 32–35.

10. J.M. Ripalda. Time reversal and negative energies in general relativity, arXiv: gr-qc/9906012 (1999).

11. See the papers listed at http://msp.warwick.ac.uk/~cpr/paradigm/

12. D.G. Saari. Mathematics and the 'dark matter' puzzle, *Am. Math. Mon.* **122** (2015) 407–423.

13. The phrase 'the exception proves the rule' is widely trotted out to dismiss awkward exceptions. I've never understood why people do this, other than as a debating ploy. It makes no sense. The word 'prove' in that context originally had the meaning 'test' – just as we still *prove* bread dough; that is, test to see if it's the right consistency. (See en.wikipedia.org/wiki/Exception_that_proves_the_rule.) The phrase goes back to ancient Rome, in the legal principle *exceptio probat regulam in casibus non exceptis* (the exception confirms the rule in cases not excepted). Which means that if your rule has exceptions, you need a different rule. That does make sense. Modern usage omits the second half, producing nonsense.

19. Outside the Universe

1. The *truly* fundamental constants are specific combinations of these quantities that don't depend on the units of measurement: 'dimensionless constants' that are pure numbers. The fine structure constant is like that. The numerical value of the speed of light does depend on the units, but we know how to convert the number if we use different units. Nothing I say depends on this distinction.

2. B. Greene. *The Hidden Reality*, Knopf, New York, 2011.

3. What matters is that there's some fixed number that's bigger than the number of states of any patch. Exact equality isn't required.

4. Numbers with huge exponents like these behave rather strangely. If you look on the web you'll find that the nearest exact copy of you is about $10^{10^{128}}$ *metres* away. I replaced that with light years, which are much bigger than metres. But actually, changing the units makes very little difference to the *exponent*, because $10^{10^{128}}$ metres is $10^{10^{128}-11}$ light years, and the exponent $10^{10^{128}-11}$ is a 129-digit number, just like 10^{128}. Their ratio is $1.000...00011$ with 125 zeros.

5. B. Greene. *The Hidden Reality*, Knopf, New York, 2011, p. 154.

6. L. Carroll. *The Hunting of the Snark*, online free at https://www.gutenberg.org/files/13/13-h/13-h.htm

7. G.F.R. Ellis. Does the multiverse really exist? *Sci. Am.* 305 (August 2011) 38–43.

8. O. Romero-Isart, M.L. Juan, R. Quidant, and J.I. Cirac. Toward quantum superposition of living organisms, *New J. Phys.* 12 (2010) 033015.

9. J. Foukzon, A.A. Potapov, and S.A. Podosenov. Schrödinger's cat paradox resolution using GRW collapse model, *Int. J. Recent Adv. Phys.* 3 (2014) 17–30.

10. Known as a 'ket' vector in Dirac's formalism for quantum mechanics. The right-hand end of a brac*ket*, OK? Mathematically, it's a vector rather than a dual vector.

11. A. Bassi, K. Lochan, S. Satin, T.P. Singh, and H. Ulbricht. Models of wave-function collapse, underlying theories, and experimental tests, *Rev. Mod. Phys.* 85 (2013) 471.

12. J. Horgan. Physicist slams cosmic theory he helped conceive, *Sci. Am.* (1 December 2014); http://blogs.scientificamerican.com/cross-check/physicist-slams-cosmic-theory-he-helped-conceive/

13. F.C. Adams. Stars in other universes: stellar structure with different fundamental constants, *J. Cosmol. Astroparticle Phys.* 08 (2008) 010.

14. V. Stenger. *The Fallacy of Fine-Tuning*, Prometheus, Amherst, 2011.

15. That is, on a log/log scale and in a specific but wide range of values, the region of parameter space for which stars can form has about one quarter the area of the whole space. This is a rough-and-ready measure, but it's comparable to what fine-tuning proponents do. The point isn't the 25%: it's that any sensible calculation of the likelihood makes it far bigger than 10^{-47}.

Epilogue

1. Adam G. Reiss and 14 others. A 2·4% determination of the local value of the Hubble constant, http://hubblesite.org/pubinfo/pdf/2016/17/pdf/pdf.

Picture Credits

The illustrations have been reproduced with the kind permission of the following:

Black and white illustrations

Atacama Large Millimeter Array, p. 33; E. Athanassoula, M. Romero-Gómez, A. Bosma & J. J. Masdemont. 'Rings and spirals in barred galaxies – II. Ring and spiral morphology', *Mon. Not. R. Astron. Soc.* 400 (2009) 1706–20, p. 183; brucegary.net/XO1/x.htm, p. 191; ESA, p. 2; M. Harsoula & C. Kalapotharakos. 'Orbital structure in N-body models of barred-spiral galaxies', *Mon. Not. RAS* 394 (2009)1605–19, p. 182 (bottom); M. Harsoula, C. Kalapotharakos & G. Contopoulos. 'Asymptotic orbits in barred spiral galaxies', *Mon. Not. RAS* 411 (2011) 1111–26, p. 180; M. Hippke & D. Angerhausen. 'A statistical search for a population of exo-Trojans in the Kepler dataset', ArXiv:1508.00427 (2015), p. 193; W. S. Koon, M. Lo, S. Ross & J. Marsden, pp. 144, 145; C. D. Murray & S. F. Dermott, *Solar System Dynamics*, (Cambridge University Press 1999), p. 90; NASA, pp. 4, 89, 98, 103, 116, 117, 139, 173 (left), 178, 197, 249; M. Proctor. Dynamo action and the Sun, *EAS Publications Series* 21 (2006) 241–73, p. 167; N. Voglis, P. Tsoutsis & C. Efthymiopoulos. 'Invariant manifolds, phase correlations of chaotic orbits and the spiral structure of galaxies', *Mon. Not. RAS* 373 (2006) 280–94, p. 182 (top); Wikimedia commons, pp. 77, 147, 152, 155, 161, 165, 173 (right), 177, 185 (right); J. Wisdom, S. J. Peale & F. Mignard. 'The chaotic rotation of Hyperion', *Icarus* 58 (1984) 137–52, p. 136; www.forestwander.com/2010/07/milky-way-galaxy-summit-lake-wv/ p. 173 (top)

Colour plates

Pl. 1 NASA/JHUAPL/SwRI; Pl. 2 NASA/JHUAPL/SwRI; Pl. 3 NASA/JPL/University of Arizona; Pl. 4 NASA/JPL/DLR; Pl. 5 NASA/JPL/Space Science Institute; Pl. 6 NASA; Pl. 7 NASA/SDO; Pl. 8 M. Lemke and C. S. Jeffery; Pl. 9 NGC; Pl. 10 Hubble Heritage Team, ESA,

Index

Italic page numbers indicate a relevant illustration only on pages without a text discussion.